DIABETES AND THE ENDOCRINE PANCREAS:
A Biochemical Approach

Diabetes
and the Endocrine Pancreas
A Biochemical Approach

William Montague

CROOM HELM
London & Canberra

British Library Cataloguing in Publication Data

Montague, William
  Diabetes and the endocrine pancreas. — (Croom Helm
  biology in medicine series)
  1. Endocrinology. 2. Metabolism
  3. Diabetes
  I. Title
  616.4'62   RC660

    ISBN 978-0-85664-888-5   ISBN 978-94-011-6379-8 (eBook)
    DOI 10.1007/978-94-011-6379-8

CONTENTS

PREFACE

This book attempts to explore the contribution that biochemistry has made, thus far, to our understanding of the endocrine pancreas and its relationship to diabetes mellitus. It was written with the aim of using an important clinical problem to illustrate, to medical students, that there are many aspects of the biochemistry taught in the early years which have direct relevance to clinical medicine. Furthermore, it is hoped that such information might provide biochemistry students with a framework on which to base further studies. To this end a selection of recent references has been placed at the end of each chapter.

In spite of considerable advances in our understanding of diabetes mellitus, it is still a disease which many physicians do not seem to comprehend. This is in part related to their lack of understanding of the molecular biology of the disease. Advances in this area have been dramatic in recent years and we are now able to offer a molecular basis for a rational approach to therapy. It may be therefore that this book will provide some physicians with the information they require to help them gain a deeper understanding of the disease.

I hope that everyone who reads this book is able to capture some of the fascination that the islets of Langerhans hold for myself and the many other workers actively engaged in trying to unravel their mysteries.

ACKNOWLEDGEMENTS

I am indebted to Simon Howell, Peter Watkins and my wife for their patient reading of the manuscript and to Simon Howell and Lelio Orci for allowing me to reproduce some of their beautiful electron micrographs. In addition, I wish to thank Pamela Broster, Amelia Dunning and Ulla Gervind-Richards for their patient typing of the manuscript.

ACKNOWLEDGEMENT

For Agnes, Claire and Lisa

1 INTRODUCTION

Diabetes mellitus is at present one of the major health problems in Europe and North America affecting at least 1 per cent of the population. Moreover, although the acute and potentially lethal metabolic derangements of diabetes can be controlled with insulin therapy, the long-term complications of the disease which may involve the cardiovascular, renal and nervous sytems reduce life expectancy by as much as a third. In spite of intensive research into the aetiology and pathogenesis of the disease many aspects of diabetes remain a mystery. However, it is now clear that diabetes is not a single disease entity, but rather a syndrome composed of a number of diseases that share glucose intolerance as a common feature. Two major types of primary diabetes mellitus are now recognised clinically, Type 1 or insulin-dependent diabetes mellitus (IDD) and Type 2 or non-insulin-dependent diabetes mellitus (NIDD). They both appear to result from a complex interaction between the individual and the environment and in some cases the outcome of this interaction may be influenced by the genetic constitution of the individual.

Diabetes mellitus is known to have affected man for many thousands of years, the earliest recorded description of the symptoms being found in the Ebers papyrus of Egypt which dates back to 1500 BC. However, it was not until the second century AD that Aretaeus of Cappadocia named the disease diabetes, the Greek word meaning 'to flow through a siphon'. He wrote 'Diabetes is a remarkable disorder. It consists of a moist and cold wasting of the flesh and limbs into urine. The disease is chronic in its character, and is slowly engendered, though the patient does not survive long when it is completely established, for the marasmus produced is rapid and death speedy'. This was a masterly description of the striking symptoms of severe diabetes, a copious flow of urine accompanied by the wasting away of both muscle and fat. In the sixth century, Hindu physicians recognised that the urine from diabetic patients tasted sweet, although it was not until the eighteenth century that the sweet-tasting substance was identified as the sugar glucose and the word mellitus, or 'honeyed', was added.

One of the first clues to the pathology underlying diabetes came in

1889 from the experimental work of Von Mering and Minkowski. They found that removal of the pancreas from dogs gave rise to a syndrome resembling diabetes, i.e. increased production of urine containing glucose. They went on to show that the pancreas was a gland of internal secretion which produced a substance that regulated glucose metabolism. Laguesse, in 1894, drew attention to the original observations of Paul Langerhans, who had in 1869 described small 'heaps' or islands of a previously unknown cell type in the pancreas. Laguesse suggested that these islands of cells should be called the islets of Langerhans, and that they were the gland of internal secretion of the pancreas. The hypothetical pancreatic secretion, the lack of which induced the diabetic state was given the name 'insuline' by De Meyer in 1900 to denote its origin from the 'insulae' of Langerhans.

Further work developed in two major directions. One consisted of extensive histological studies of the islets which led to the finding of several distinct cell types and foreshadowed our present knowledge that the islets produce and secrete several different hormones. The other pathway consisted of the numerous attempts to extract from the pancreas a potent, antidiabetic material, the insuline of De Meyer. Many technical problems had to be overcome before insulin could be extracted from the pancreas in sufficient quantities and pure enough for human use. With the benefit of hind-sight, we now know that most of the problems were due to the facts that insulin is a protein and the pancreas is rich in proteolytic enzymes. This meant that steps had to be taken to minimise proteolytic destruction during extraction. Moreover, any material isolated could not be given orally to test its antidiabetic activity as it would be degraded by the digestive enzymes. Banting and Best successfully overcame these problems and in 1921 produced the first useful and consistently successful insulin preparation for the treatment of diabetes. The discovery of insulin was hailed as a cure for diabetes because it lowered blood-glucose levels, controlled the acute symptoms of the disease and prevented the coma and death that sometimes came within days of the onset of symptoms.

However, it became apparent several years later that diabetics who had been on insulin for a long time were found to have an unusually high incidence of cardiovascular disease, renal disease, gangrene and blindness. Insulin treatment thus controlled the early metabolic symptoms of diabetes, but in some cases not the development of long-term complications. These observations raised the question as to whether the long-term complications of diabetes are the result of inadequate or inappropriate therapy or whether they represent a generalised tissue defect which manifests itself initially in the islets of Langerhans and subsequently in other tissues. Numerous studies have therefore been undertaken with the objective of defining, in detail, the relationship between the islets of Langerhans and diabetes with its complications. It is the purpose of this book to outline the current status of our knowledge concerning diabetes and the endocrine pancreas.

2 THE STRUCTURE OF THE ENDOCRINE PANCREAS

In man the pancreas is an organ of both exocrine and endocrine function. The adult organ is an elongated structure that can be divided into three regions, the head, the body and the tail. It develops from two outgrowths (diverticula) of the embryonic duodenum, one of which forms most of the head region while the other forms the remainder of the organ. These endodermal outgrowths fuse to form a typical exocrine gland connected by ducts to the gut.

The endocrine function of the pancreas is performed by the islets of Langerhans which comprise only 1-2 per cent of the wet weight of the adult pancreas. The islets, named after their discoverer Paul Langerhans, are clusters of cells found scattered throughout the pancreas. They are produced from cells which bud off from the embryonic duct system and divide to form the isolated clusters of cells. The clusters lose their connections with the duct system after 2-3 months *in utero*, at about the time that hormone production begins. The adult human pancreas contains approximately one million islets distributed unevenly in the pancreas, there being a larger number in the tail than in the body and head of the pancreas. They are round to ovoid structures (Figure 2.1) containing small, polygonal endocrine cells (Figure 2.2) arranged as cords along extensive capillary channels. The islets in man vary in size (20-300 μm in diameter) due largely to differences in the numbers (2,000-6,000) and types of cells in each islet. They are highly vascularised and extensively innervated structures, features which relate to the fact that their activity is regulated both by the autonomic nervous sytem and by factors in the circulation.

Vascular Structure of the Islets

The pancreas, and hence the islets, receives its arterial blood supply from the splenic, hepatic and superior mesenteric arteries and the venous drainage is into the splenic and superior mesenteric veins. The islets are highly vascularised with an elaborate network of anastomosing capillaries within the islet. The capillaries are composed of fenestrated

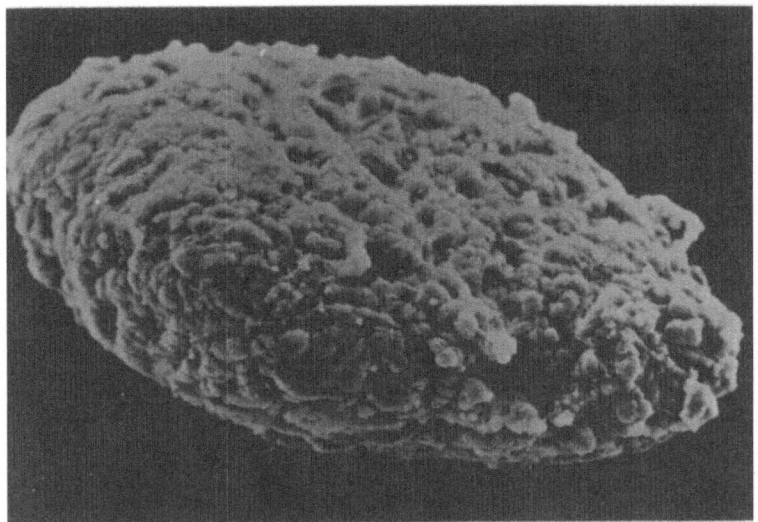

Figure 2.1: Scanning Electron Micrograph of a Single Islet of Langerhans. The individual cells of the islet can be clearly seen. Magnification approx. 1000 X. Courtesy of S.L. Howell, London.

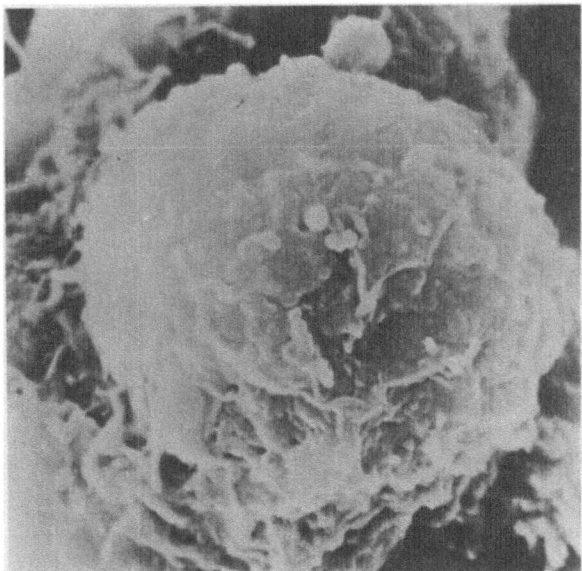

Figure 2.2: Scanning Electron Micrograph of an Individual Islet Cell. Magnification approx. 10,000 X. Courtesy of S.L. Howell, London.

endothelial cells, a feature typical of capillaries in other endocrine glands. It is thought that the fenestrations, areas of fused plasma membrane, may facilitate the rapid exchange of material across the endothelium. Each islet cell is close to a capillary and this should allow the rapid transport into the vascular space of hormones released from the endocrine cells. However, the islets are surrounded by a basement membrane separate to the capillary basement membrane and before entering the bloodstream, hormonal products must traverse both these membranes as well as the endothelial cells of the capillary wall.

Innervation of the Islets

The pancreas is innervated by the autonomic nervous system with nerve fibres from the vagus nerve (parasympathetic) and the greater and middle splanchnic nerves (sympathetic). Some of these fibres terminate at the periphery of the islets and some actually within the islet and adrenergic and cholinergic terminals have been identified at both locations. The autonomic nervous system plays an important role in modulating the secretory response of the islet cells and this will be considered further in Chapter 3.

Islet Cell Types

A variety of light and electron microscopic techniques have been used to establish that there are several distinct endocrine cell types within the islet, each cell type containing a different hormone stored within secretory granules. The major cell types are B-cells which contain insulin, A-cells which contain glucagon, D-cells which contain somatostatin and the pancreatic polypeptide containing cells (PP-cells). The relative proportions of the cell types differ in islets in different parts of the human pancreas. All of the islets are normally composed of 60-70 per cent B-cells and 5-10 per cent D-cells. However, islets located at the head of the pancreas contain, in addition, 20-25 per cent A-cells and 5 per cent PP-cells whereas islets located at the tail of the pancreas contain 15-20 per cent PP-cells and only 5 per cent A-cells. The physiological significance of this difference is unknown, although it may reflect the different embryological origins of these two regions.

The various cell types can be distinguished on the basis of the ultrastructural appearance of their secretory granules (Figure 2.3). The

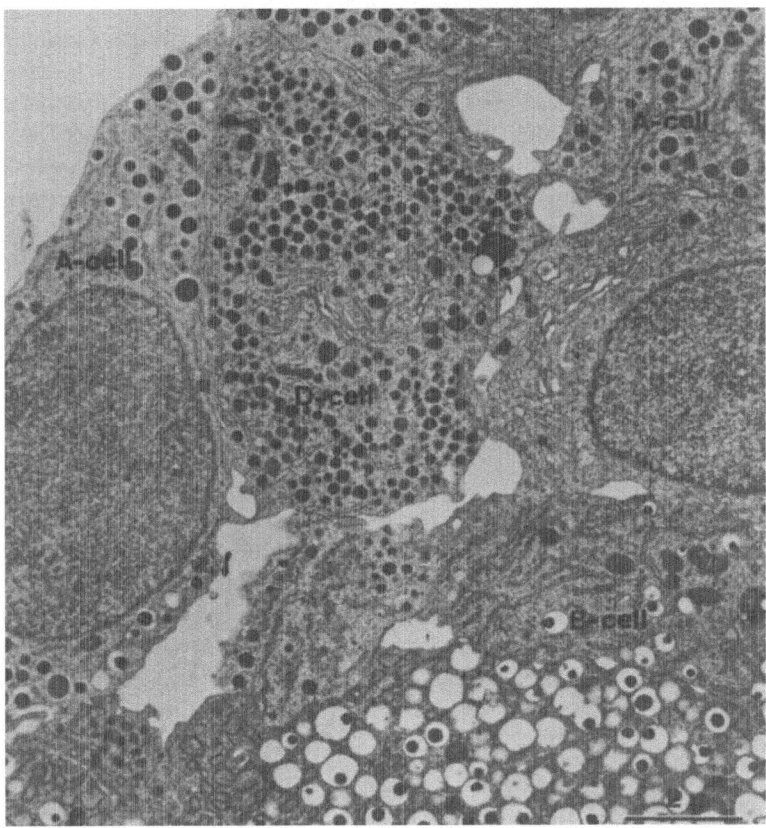

Figure 2.3: Electron Micrograph of a Section through an Islet of Langerhans Isolated from the Rat Pancreas. The section shows at the electron microscopical level the differences between the secretory granules of a B-cell, an A-cell and a D-cell. A fourth cell type, the pancreatic polypeptide cell (PP-cell) is not shown in this section. Magnification 9,000 X. Courtesy of L. Orci, Geneva, Switzerland.

B-granules have a large space separating the electron-dense granule contents from the encasing membrane. Moreover the granule contents have rectangular profiles with a crystalline matrix containing lines of repeating periodicity of approximately 50Å. The A-granules are smaller than B-granules and the granule membrane is closely applied to the extremely electron-dense granule core, there being only a small less dense area between the two. The D-granules are larger (250-450 nm) than A and B-granules and the granule membrane is closely applied to the granule content which has a low electron density. The pancreatic polypeptide granules have an electron density which is intermediate between that of the A and D-cells and they show a marked variability in profile.

The four cell types bear a constant topographical relationship to one another, although the particular cellular arrangements may differ from species to species. In man, islets located at the head of the pancreas contain an outer rim composed of A- and PP-cells. The D-cells are situated immediately under this outer cell rim in close association with the A-cells and the B-cells which form a central mass in the islet. The islet thus contains a heterocellular cortical region in which there are local contacts between different cell types. The function of the heterocellular region may involve local interactions of the various secretory products of neighbouring islet cells on one another, so-called 'paracrine' effects. These effects may be important in ensuring an integrated secretory response of the various islet cells.

In spite of differences in hormonal contents and secretory granule structure, the cell types show certain common ultrastructural features which are related to their common functions of polypeptide hormone synthesis, storage and secretion. The most extensively studied islet cell type is the B-cell and its structure may, in general terms, be regarded as being typical of the other islet cells. Furthermore, it is generally assumed that the results of biochemical studies also reflect the B-cell although they do in fact represent the sum total of the different islet cells.

General Ultrastructural Features of Islet Cells

Each islet cell is surrounded by a distinct plasma membrane which has a thickness of 7.5 nm and shows the classical trilaminar 'unit' structure. The cells have a polyhedral shape with a mean diameter in the region of 10 μm and a surface area of approximately 1,000 μm^2.

On the outer surface of the plasma membrane is the cell coat, a layer of variable thickness which probably represents the extended glycosidic residues of membrane glycoproteins and glycolipids. The cell coat appears to be involved in cell surface events such as cellular recognition and cellular adhesion and many cell surface receptors are thought to be localised in this layer.

On the inner surface of the plasma membrane of the B-cell is another specialised layer known as the cell web (Figure 2.4). This layer is composed principally of thin (4-6 nm) filaments (microfilaments) running parallel to the plasma membrane. The filaments are polymerised structures composed largely of the subunit protein actin. The total actin

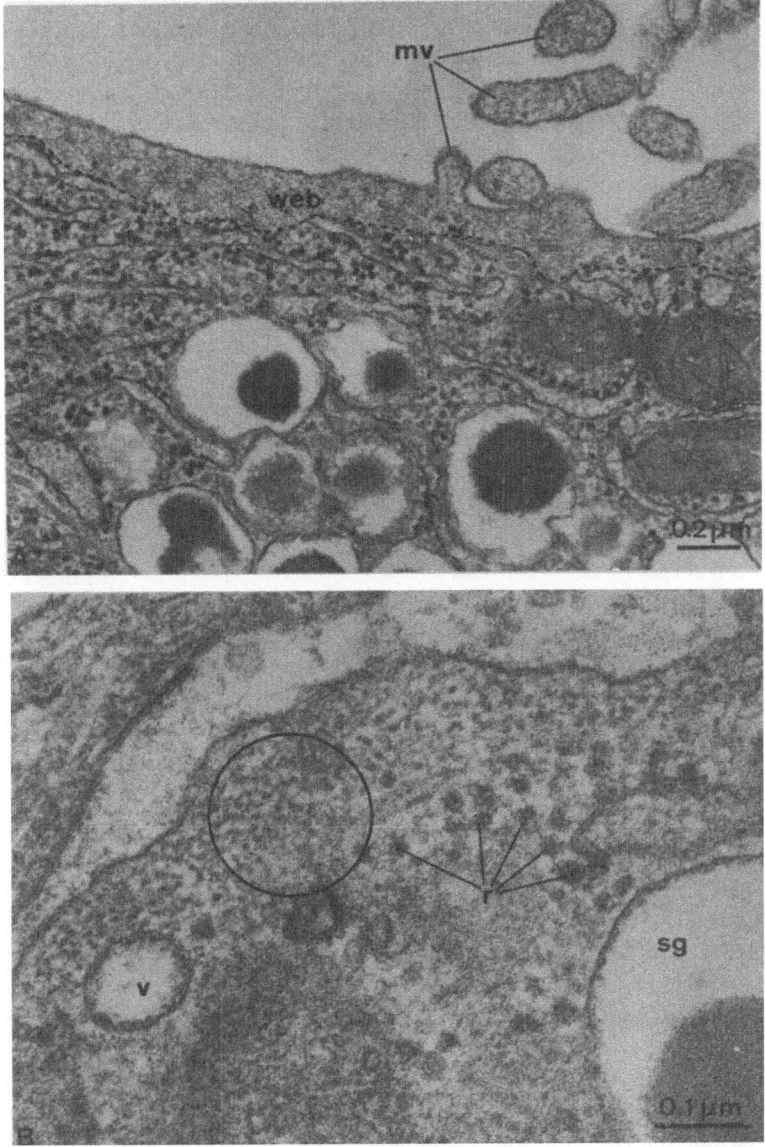

Figure 2.4: Electron Micrographs of a Section through the B-cell of a Rat Islet of Langerhans. (A) The cell web (microfilaments) is present just beneath the plasma membrane and extends into the core of microvillous processes (mv). X 54,000. (B) At higher magnification, the network of microfilaments of the cell web are more clearly demonstrated. v : microvesicle; r: ribosomes; sg: secretory granule. X 166,000. Courtesy of L. Orci, Geneva, Switzerland.

content of islet cells represents 1-2 per cent of the total islet tissue protein and in the unstimulated islet cell 20-30 per cent of the actin is polymerised to form microfilaments. During secretion this proportion can increase to 60-70 per cent, suggesting that microfilaments may play an important role in the secretory process. In addition the microfilaments of the cell web are thought to be responsible, at least in part, for the active movement of the plasma membrane (ruffling).

At several points on the cell periphery regions of the plasma membranes of adjacent islet cells may be specialised to form intercellular contacts. These highly specialised areas of membrane can be formed between any one of the various islet cell types and can take the form of desmosomes, tight junctions or gap junctions.

Desmosomes consist of focal condensations of extracellular material between the two membranes with adjacent dense cytoplasmic plaques on the inner surfaces of the two membranes. They are thought to represent sites of cell-to-cell adhesion and help to maintain the structural integrity of the islet as a whole.

Tight junctions are formed when the outer leaflets of the plasma membranes of two adjacent cells fuse. These areas of fusion may extend linearly and form networks which enclose areas of extracellular space between adjacent cells. The extent of such enclosures can be varied as the junctions can readily be assembled and disassembled. In this way islet cells may have the capacity selectively to open or close certain plasma membrane areas to substances in the extracellular fluid. Such compartmentalisation may also guide secretory products directly to the circulation via the perivascular space for endocrine interaction, or to receptors on the surface of adjacent cells for paracrine interaction.

Gap junctions are formed when the outer leaflets of the plasma membranes of two adjacent cells come close together along a narrow (2-4 nm) gap of intercellular space. They consist of focal aggregates of a number (10-20) of closely packed intramembrane particles. The number of gap junctions as well as their size appears to be related to the functional activity of the cell and in the resting B-cell it is in the region of 20-30 per 100 μm^2 of membrane. This number increases during extended periods of secretory activity. Many of the intramembrane particles of the gap junction have small central pits which may correspond to the hydrophilic channels which appear to bridge the cytoplasmic compartments of the two cells sharing the junction. These channels are thought to be responsible for the exchange of ions and small molecules which occurs across gap junctions and which makes possible electrical and metabolic coupling of adjacent cells. Such intracellular communication between the various islet cell types may be important in ensuring that the individual islet cells respond in a synchronised manner and that the integrated response of the islet as a whole is appropriate to the needs of the organism.

Inside the islet cell plasma membrane the cytoplasm is divided by intracellular membranes into a large number of specific compartments. These include the nucleus, mitochondria, the rough and smooth endoplasmic reticulum, the Golgi apparatus and vesicles of various sizes, including transfer microvesicles and secretory granules (Figures 2.5 and 3.3a). There are, however, relatively few lysosomes in the normal islet cell. The intracellular membranes like the plasma membrane show the classical trilaminar 'unit' structure. The membranes limiting the endo-

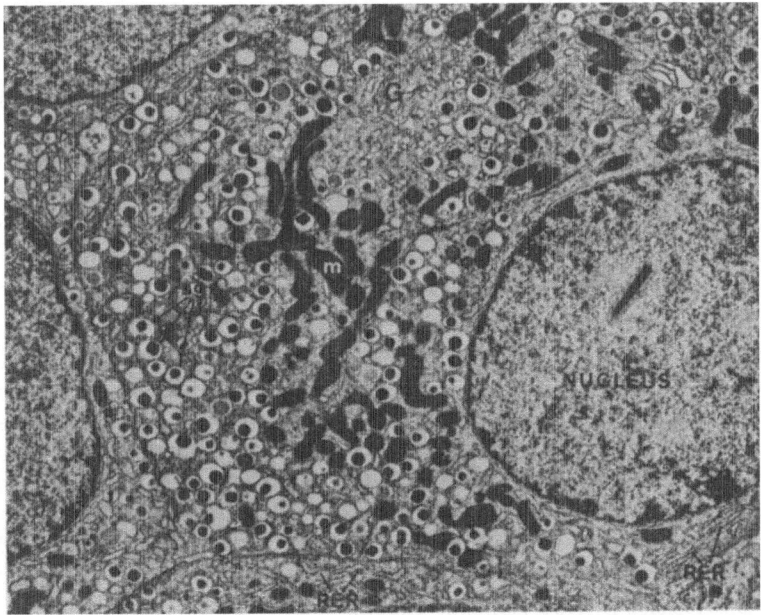

Figure 2.5: Electron Micrograph of a Section through the B-cell of a Rat Islet of Langerhans. The section shows the following subcellular compartments, G, Golgi apparatus; sg, secretory granules; m, mitochondria; RER, rough endoplasmic reticulum and NUCLEUS. Magnification 10,000X. Courtesy of L. Orci, Geneva, Switzerland.

plasmic reticulum, the transfer microvesicles and the outer Golgi cisternae are however thinner (~6 nm) than those limiting the inner Golgi stacks and the secretory granules (~7.5 nm).

Each islet cell contains approximately 10 pg of DNA localised in a single nucleus which occupies in the region of 12 per cent of the cell volume. This is similar to the total volume occupied by the 13,000 or so secretory granules found in a normal B-cell. These granules are membranous sacs with a diameter of 0.3 μm which contain a more or less dense core of stored polypeptide hormone ready for secretion. Their absolute size, the extent to which the core fills the sac and the density of the core varies between islet cell types but is relatively constant within a particular type. In the non-stimulated cell the granules appear to be distributed generally throughout the cytoplasm with no apparent concentration in any region.

The endoplasmic reticulum is composed of flattened cisternae with a narrow lumen and occupies about 20 per cent of the cell volume. Ribosomes are attached to the outer surface of between 60-70 per cent of the endoplasmic reticulum giving a ratio of rough to smooth endoplasmic reticulum of approximately two to one. In the non-stimulated B-cell it has been estimated that at least one million ribosomes are located in association with the surface of the rough endoplasmic reticulum and this number is likely to increase dramatically when hormone synthesis is stimulated.

The Golgi apparatus is usually located near the nucleus and is composed of stacked smooth-walled cisternae with a rather wide clear lumen and associated vesicles. On one of its sides, the outer 'forming' surface, the Golgi area is in topographical relationship with the endoplasmic reticulum and on the other, the inner 'maturing' surface, with the newly formed secretory granules. This topographical relationship is a reflection of the functional relationship of these various compartments since material synthesised on the rough endoplasmic reticulum and destined for secretion, is transported to the outer Golgi cisternae in microvesicles derived from the endoplasmic reticulum. This material is subsequently packaged into vesicles derived from the inner Golgi cisternae and these vesicles ultimately mature into the secretory granules. The size of the Golgi complex varies with the functional status of the individual cell. When stimulated the islet cells are amongst the most metabolically active cells in the body, as the synthesis and secretion of protein are both energy-requiring processes. This intense metabolic activity is reflected by the large number (approximately 1,000) of mitochondria in an islet cell. They are relatively small cylindrical structures which occupy approximately 4 per cent of the islet cell volume.

In addition to these various membrane-limited compartments, there are extensive networks of microtubules and microfilaments in islet cells (Figure 2.6). Unlike the microfilamentous system which appears to be confined to the cell web, the microtubule system is present in all parts of the cell and it may be organised into a network which radiates from the nucleus towards the plasma membrane. Microtubules are long, cylindrical structures with an outside diameter of 25 nm. They are highly polymerised structures composed largely of a single subunit protein known as tubulin. This is a 110,000 molecular weight dimeric protein consisting of two almost identical molecules, α- and β-tubulin, each with a molecular weight close to 55,000. Polymerised microtubules in islet cells are in dynamic equilibrium with a pool of tubulin subunits and changes in this equilibrium have dramatic effects on the size of the microtubule network. In the resting B-cell 20-30 per cent of the total tubulin is in the form of polymerised microtubules and the total length of the microtubule network has been estimated at 1,300 μm. During periods of active secretion there is a shift in the equilibrium between

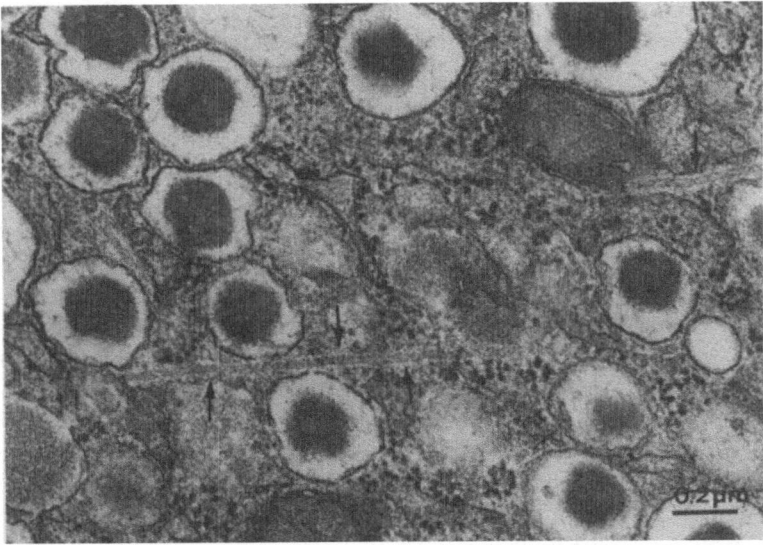

Figure 2.6: Electron Micrograph of a Rat B-cell Stained to Show Microtubules. The arrows indicate microtubules. Magnification 59,000X. Courtesy of L. Orci, Geneva, Switzerland.

subunits and polymerised microtubules such that 40-50 per cent of the tubulin becomes polymerised and this would be expected to approximately double the size of the microtubule network. The control of this equilibrium is an important aspect of the secretory process since microtubules are known to play an important role in hormone secretion from islet cells.

Ultrastructural Aspects of Secretion

The secretion of hormones from the various islet cell types occurs via a common mechanism which involves two stages. Initially the secretory granules are mobilised from storage in the general cytoplasm of the cell and directed towards the plasma membrane. This stage is known as margination and is thought to involve the participation of the cell's microtubule/microfilamentous system. Once the granules are at the plasma membrane the granule contents can be released into the extracellular space by exocytosis (Figure 2.7). This final stage involves the fusion of the secretory granule membrane with the cell membrane and exposes the interior of the granule to the extracellular fluid. The granule contents dissolve in the fluid and diffuse away from the cell.

As a result of exocytosis segments of the granule membrane become incorporated into the plasma membrane and this should lead to an increase in islet cell surface during periods of active secretion. Indeed, it has been estimated that during a sustained period of secretion lasting for one hour an area of granule membrane equal to that of the plasma membrane fuses with the plasma membrane. However, the cell surface does not significantly enlarge during sustained secretion and there must therefore be some mechanism to dispose of excess membrane. This is achieved by the process of endocytosis which involves the internalisation of segments of the plasma membrane at a rate equivalent to that at which the granule membrane is added during exocytosis. Endocytosis involves the invagination of areas of plasma membrane. The invaginations are subsequently pinched off to form microvesicles which eventually become incorporated into the membranes of the Golgi area. It is not known whether areas of membrane derived from the secretory granules are specifically retrieved from the cell surface during endocytosis or whether the process is relatively non-selective. Various studies have, however, revealed marked differences between the granule and plasma membrane. Thus freeze-fracture studies have shown that the plasma membrane is relatively rich in intramembranous particles

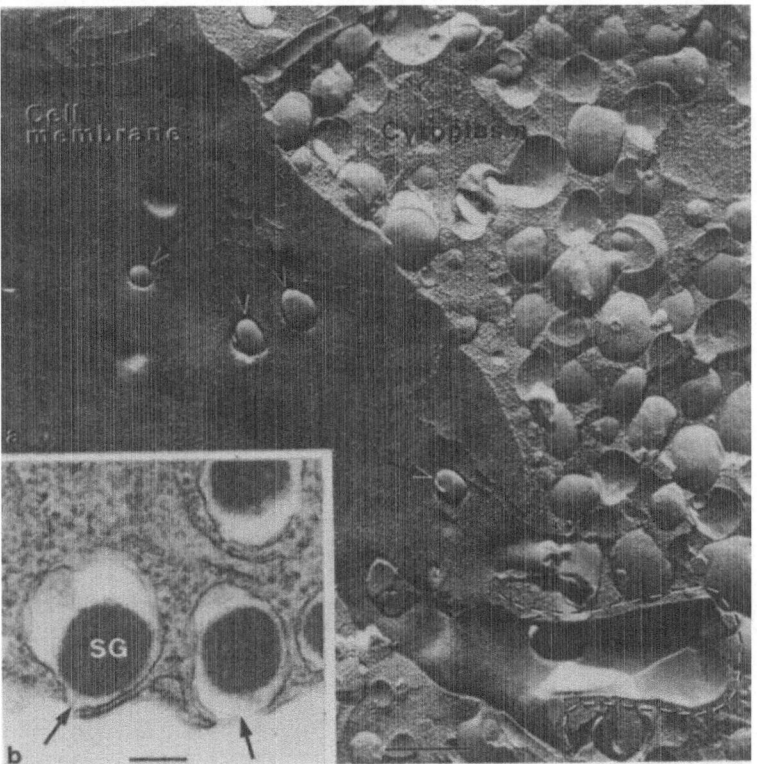

Figure 2.7: Electron Micrographs of B-cells During Insulin Secretion. (a) Freeze-fracture replica of an islet cell stimulated with glucose and showing several exocytotic figures (arrows) on the P-face of the cell (plasma) membrane. The exocytotic figures represent stages at which the granule core is being extruded after the fusion of the granule limiting membrane with the plasma membrane. In the cross-fractured cytoplasm, one sees numerous profiles of secretory granule limiting membranes as well as a large concavity (outlined by a dotted line) merging with the cell membrane. Note that the membrane limiting the concavity (probably representing the membrane of several secretory granules fused together in a so-called chain release) is very smooth as compared with the richly particulated P-face of the cell membrane. (b) Typical image of secretory granules (SG) undergoing exocytosis in a B-cell. In this thin section, the arrows point to the opening of the inside of the granule in the extracellular space which occurs through fusion of the granule's limiting membrane with the cell (plasma) membrane. Magnification: (a) × 14,000; (b) × 48,000. The horizontal bar represents (a) 1 μm and (b) 0.2 μm. Courtesy of L. Orci, Geneva, Switzerland.

$(2,000/\mu m^2)$ while the mature granule membrane is relatively poor $(200/\mu m^2)$. These intramembranous particles vary in size between 8 and 10 nm and are thought to represent aggregations of membrane proteins. In addition, the mature granule membrane appears to be relatively rich in cholesterol while the plasma membrane is relatively poor. When the granule membrane approaches the plasma membrane prior to exocytosis, a particle-free and cholesterol-rich area develops in a limited zone of the plasma membrane that overlies the secretory granule. After fusion has taken place the region devoid of particles but rich in cholesterol persists around the exocytotic opening. These differences in granule and plasma membrane structure may well provide a molecular basis for a relatively selective endocytotic process.

Further Reading

Dean, P.M. Ultrastructural Morphometry of the Pancreatic β-Cell. *Diabetologia* (1973) *9*, 115-119

Howell, S.L. & Tyhurst, M. Microtubules, Microfilaments and Insulin Secretion. *Diabetologia* (1982) *22*, 301-308

Orci, L. A Portrait of the Pancreatic β-Cell. *Diabetologia* (1974) *10*, 163-187

Orci, L. Macro- and Micro-Domains in the Endocrine Pancreas. *Diabetes* (1982) *31*, 538-565

Orci, L. Membrane Cycling in Secretion. *Current Topics in Cellular Regulation* (1981) *18*, 531-550

3 INSULIN SYNTHESIS, STORAGE AND SECRETION

Introduction

The identification of insulin as the antidiabetic hormone by Banting and Best in the early 1920s was not only a milestone in our understanding and treatment of diabetes mellitus, but it also provided a powerful stimulus to the development of techniques for the isolation and characterisation of the biologically active molecule. Insulin was first crystallised in 1926 by J.J. Abel and at that time there was great controversy as to whether the crystals he had obtained were entirely the active hormone or merely the vehicle for a smaller active moiety. Such controversy seems remarkable in the light of our present detailed knowledge of the structure and properties of insulin. It should be remembered, however, that many of the modern techniques of protein chemistry were developed through the clinical need for insulin and that this made it abundantly available for the chemist, biochemist and the crystallographer to study. These studies have contributed to our knowledge of intermediary metabolism, the synthesis, structure and function of proteins and many other fields of biology.

Structure

The determination of the primary structure of bovine insulin by Sanger and his associates in 1955 provided the first known protein structure. This demonstration laid to rest the notion that proteins were not defined chemical entities and it provided an important corner-stone of molecular biology which subsequently led to the recognition of the existence of the genetic code.

The amino acid sequences of many insulins, from the primitive hagfish to man have been determined, and the three-dimensional structure of several insulins has been established by X-ray crystallography. Compared to many other proteins, insulin has been highly conserved in evolution with an amino acid substitution rate of about 1×10^{-9}/locus/year. Thus, the insulin of a very primitive vertebrate, such as the hagfish,

27

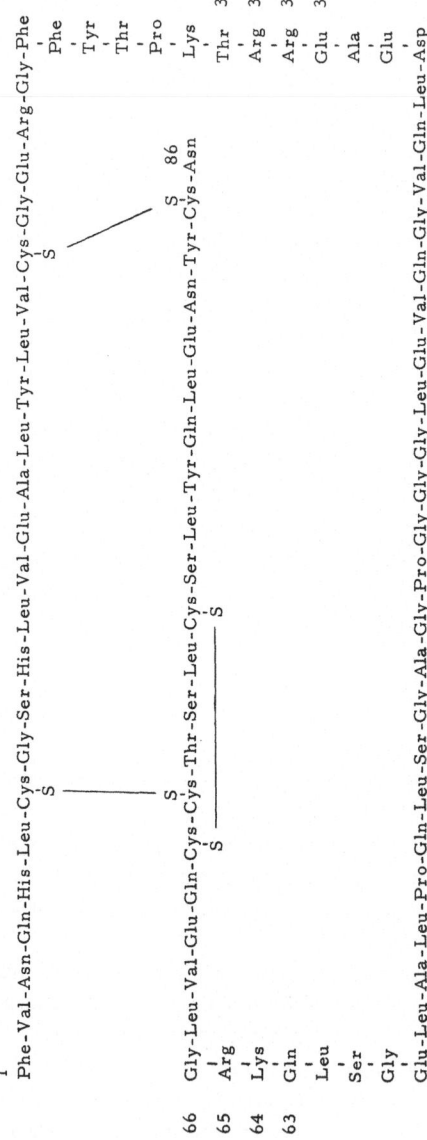

Figure 3.1: The Primary Structure of Human Insulin and Proinsulin. 1-30 = B-chain; 31-65 = Connecting peptide; 33-63 = C-peptide; 66-86 = A-chain; 31, 32, 64 and 65 = Amino acids released during conversion of proinsulin to insulin.

differs from that of man in only about 38 per cent of its residues. It is thought that those regions of the molecule that are highly conserved may play important roles in maintaining structural features needed for biological activity.

The human insulin molecule has a molecular weight of approximately 6,000 daltons and contains 51 amino acids which are arranged as two polypeptide chains (Figure 3.1). The A chain contains 21 amino acids and is linked to the B chain by two disulphide bridges. There is in addition an interchain disulphide bridge on the A chain.

Insulin is normally present in the circulation at a concentration in the region of 10^{-10} molar and at this concentration it is transported in solution as a monomer. At higher concentrations insulin readily forms dimers and these aggregate in the presence of zinc to form hexamers, indeed insulin is stored in the B-granules as a crystalline hexamer complex with two atoms of zinc per hexamer.

The elucidation by Hodgkin and co-workers of the three-dimensional structure of insulin represented an important step in our understanding of the insulin molecule since knowledge of the spatial organisation of the molecule should provide information about the molecular mechanism of binding and action of insulin. In the crystalline state insulin has many of the features of a globular protein and it probably circulates in the blood as a relatively stable globular structure. A schematic representation of the three-dimensional structure of insulin is shown in Figure 3.2. The B chain is a U-shaped structure with a single a-helical region towards the amino-terminus, while the A chain contains two regions of a-helix and lies within the hydrophobic cleft formed by the B chain.

Various studies have indicated that residues 23-26 of the carboxy-terminal region of the B chain are part of the active site of the hormone. Particularly relevant to this suggestion are the findings that these residues (-Gly-Phe-Phe-Tyr-) are invariant, that the biological activity of insulin decreases progressively as these residues are sequentially removed and that synthetic peptides containing residues B_{22-26} have some insulin-like activity. Mutations which affect the amino acid composition of this region would therefore be expected to have a profound effect on activity. Indeed, a patient has been described who synthesised an abnormal insulin molecule in which one of the phenylalanine residues in this region was replaced by a leucine. Not surprisingly this modified protein has only 10 per cent of the biological activity of normal human insulin.

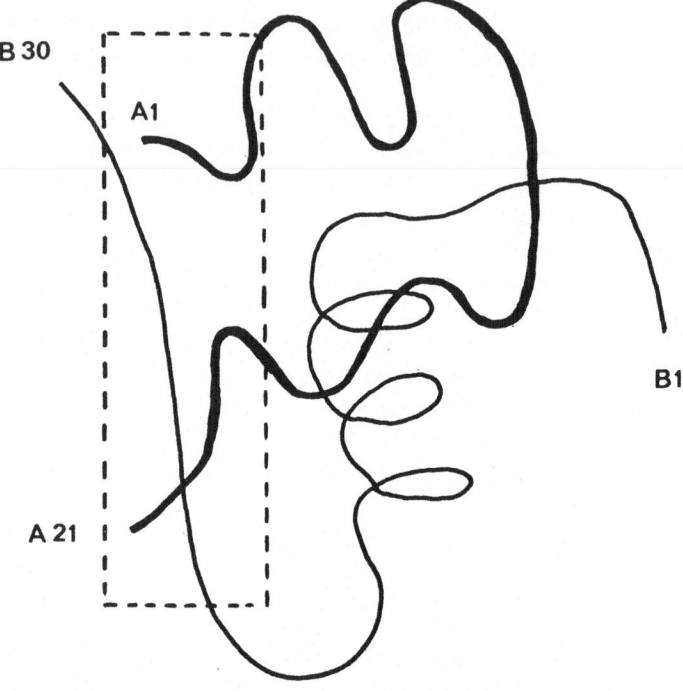

Figure 3.2: Probable Conformation of Human Insulin in Solution.
⌐‾ ‾ ‾ ‾¬ Indicates the region of the áctive centre of the molecule.

Synthesis

Prior to 1967 it was generally believed that insulin was formed by the combination of separately synthesised A and B chains. However, the discovery by Steiner and his associates of proinsulin, a relatively inactive precursor molecule which contains the A and B chains of insulin within a single polypeptide chain, cast doubt on this hypothesis and it was subsequently shown that proinsulin is indeed a biosynthetic precursor of insulin which is converted into the smaller active insulin molecule by proteolytic cleavage of specific peptide bonds. The mechanism of insulin biosynthesis via proinsulin has served as a model for the study of many other hormonal systems and it is now known that the other pancreatic hormones (glucagon, somatostatin and pancreatic polypeptide) are also produced by proteolytic cleavage of larger precursor molecules. Insulin is synthesised by way of a complex pathway (Figures 3.3a and

3.3b) which involves several cellular organelles and at least two inter-mediates (preproinsulin and proinsulin). The complexity of the path-way appears to be necessary to ensure that the molecule is synthesised with the correct orientation of the disulphide bridges and that it is packaged within the B-granules ready for secretion.

Advances in the isolation and the *in vitro* translation of mRNA have made it possible to determine the initial translation product of the mRNA for a number of polypeptide hormones and to follow their sub-sequent conversion. The mRNA for insulin has been isolated from a num-ber of species and shown to consist of approximately 600 nucleotides, with a molecular weight of 210,000 daltons (Figure 3.4). It has a poly-adenylate region at the $3'$ end and is capped with a 7-methylguanosine residue at its $5'$ end, features which are common to other eukaryotic mRNA molecules. Translation of this mRNA in a wheat germ cell free translation system gives rise initially to a single chain polypeptide (preproinsulin) having a molecular weight of 11,500 daltons and con-taining 109 amino acids. This molecule consists of a proinsulin molecule with an additional hydrophobic extension of 23 amino acids (pre-region) on the *N*-terminal region of the proinsulin B chain. The hydrophobic region is very rapidly removed *in vivo*, probably before the peptide chain is completed.

Preproinsulin is one of a family of rapidly cleaved precursor forms of secreted proteins which have been identified in several tissues. The hydrophobic terminal extensions of these precursors may promote the association of ribosomes with the membrane of the endoplasmic reticu-lum, thereby leading to the vectorial discharge of the nascent secretory peptide across the membrane into the cisternal space. This 'signal hypothesis' appears to provide a simple and plausible explanation for the initial sequestration of secretory proteins during their biosynthesis and for the formation of the rough endoplasmic reticulum.

Regulation of Insulin Biosynthesis

The rate of insulin biosynthesis by the B-cell appears to be largely under the control of the extracellular glucose concentration. At normal fasting blood glucose concentrations there is a low basal rate of syn-thesis. However, when the blood glucose concentration increases above the fasting level there is a rapid (1-2 min) and dramatic increase in the rate of proinsulin synthesis. There is a sigmoid relationship between glucose concentration and the rate of proinsulin synthesis with a

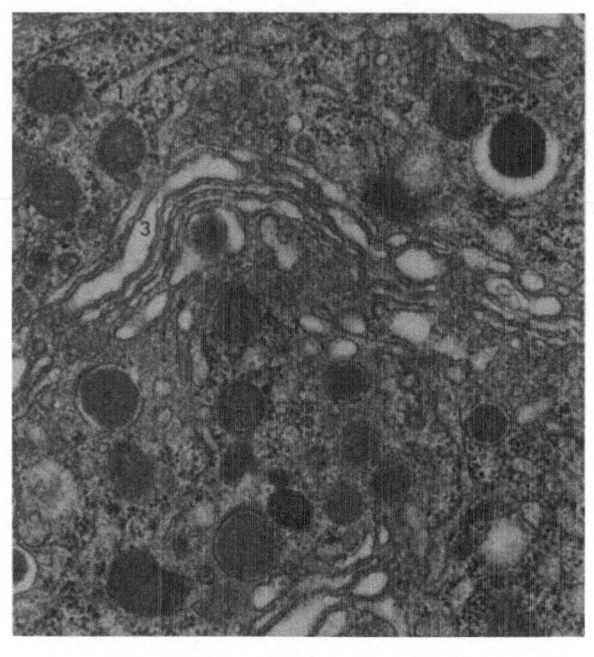

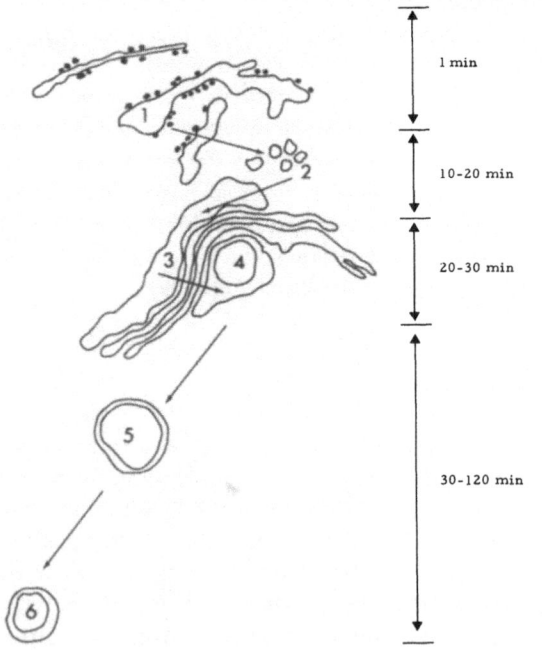

threshold of 2-4 mM glucose. This is lower than the threshold for glucose-stimulated insulin secretion (4-6 mM) and ensures that adequate stores of insulin are available in the B-cell for secretion when required. The molecular basis for the effect of glucose on insulin biosynthesis is not known in detail, although it does appear to involve transcriptional, translational and post-translational events.

In addition to glucose many other agents are able to stimulate pro-insulin biosynthesis (Table 3.1), although it is unlikely that they play an important physiological role. All of these agents also stimulate insulin secretion and this has given rise to speculation that regulation of the two processes may be linked. However, there is much evidence to suggest that the two are not obligatorily linked (Table 3.2).

Although the extracellular glucose concentration appears to determine the rate of insulin synthesis in the short term, there is evidence that the rate can be modulated in the long term by changes in the physiological status of the animal. Thus, during periods of starvation the basal rate of synthesis falls, whilst during the latter stages of pregnancy the rate increases. It is not known what determines these adaptive changes of the B-cell.

Biosynthetic Organisation of the B-cell

The B-cell shares many ultrastructural features with other cells that synthesise and store secretory proteins. These include an extensive rough endoplasmic reticulum and Golgi apparatus, and numerous mitochondria and secretory granules. Insulin, in common with other exportable

◁ Figure 3.3a: Electron Micrograph of a Portion of a B-cell showing the Subcellular Components Involved in the Synthesis and Storage of Insulin. The following subcellular components can be identified (1) rough endoplasmic reticulum studded with ribosomes, (2) transfer microvesicles, (3) Golgi apparatus, (4 and 5) immature storage granules, (6) mature storage granule. Magnification X 59,000. Courtesy of L. Orci, Geneva, Switzerland.

◁ Figure 3.3b: Diagrammatic Representation of the Events Involved in the Synthesis and Storage of Insulin. This diagram is based on Figure 3.3a. Preproinsulin is synthesised on ribosomes associated with the rough endoplasmic reticulum (1) and converted into proinsulin within the rough endoplasmic reticulum. The proinsulin is then transported in transfer microvesicles (2) to the Golgi apparatus (3) where it is packaged into storage granules (4). Maturation of the storage granules (5) is accompanied by the conversion of proinsulin into insulin and the mature storage granule (6) contains insulin and C-peptide (6). The time scale of these events is shown to the right of the diagram.

PREPROINSULIN mRNA

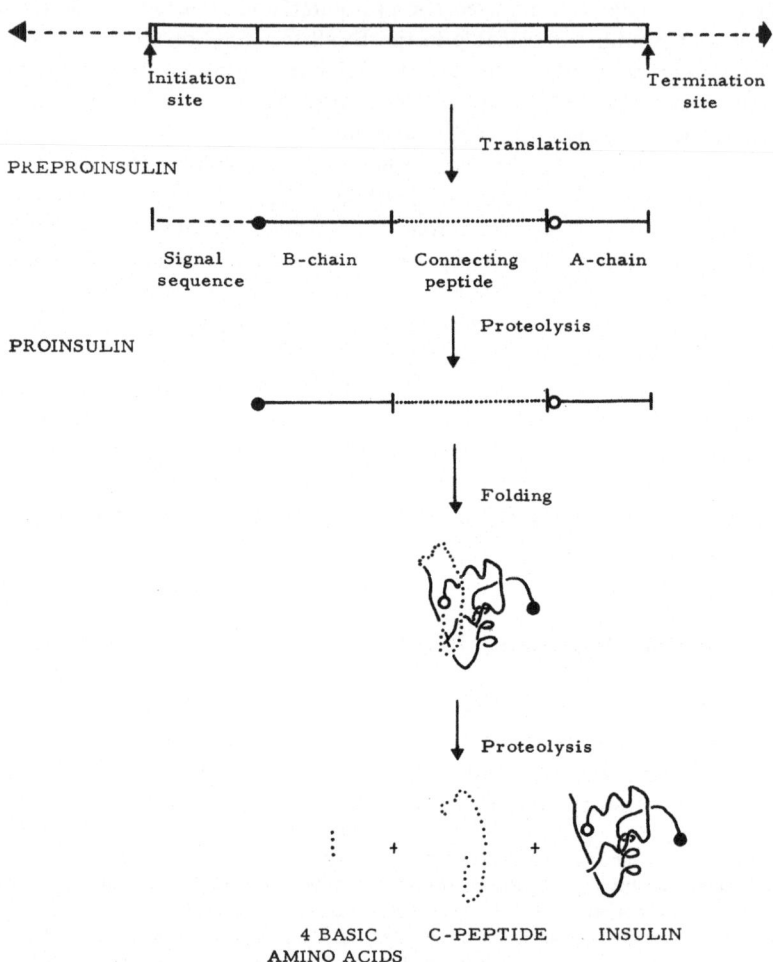

Figure 3.4: Translation of Preproinsulin mRNA in the B-cell

Table 3.1: Agents which Affect Insulin Biosynthesis

Stimuli	Inhibitors	No Effect
Glucose	Adrenaline	Galactose
Mannose		Sulphonylureas
Dihydroxyacetone		
Glyceraldehyde		
Leucine		
N-Acetylglucosamine		
α-Ketoisocaproate		
Glucagon		
Methylxanthines		

Table 3.2: Differences between the Regulation of Insulin Biosynthesis and Insulin Secretion

	Insulin synthesis	Insulin secretion
Ca^{2+} dependency	No	Yes
Mg^{2+} dependency	Yes	No
Effect of cycloheximide	Inhibits	None
Effect of puromycin	Inhibits	None
Effect of trifluoperazine	None	Inhibits
Effect of sulphonylureas	None	Stimulates
Glucose sensitivity (threshold concentration)	2-4mM	4-6mM

proteins, is synthesised by ribosomes associated with the rough endoplasmic reticulum. Translation of the insulin mRNA occurs on polyribosomes consisting of 5 to 7 ribosomes and initially produces the preregion of the preproinsulin molecule. The hydrophobic nature of the preregion promotes its interaction with the lipid-rich endoplasmic reticulum membrane and ensures the vectorial discharge of the preproinsulin through the endoplasmic reticulum membrane into the cisternal space. In the cisternal space preproinsulin rapidly undergoes proteolytic cleavage to produce proinsulin which itself undergoes rapid peptide chain folding and sulphydryl oxidation. The folded proinsulin is then transported to the Golgi apparatus in a process that takes about 20 minutes and involves microvesicles (Figure 3.3b).

The Golgi apparatus is the site where the proinsulin is packaged into secretory granules, the limiting membranes of which are formed by evagination of the Golgi membranes. The conversion of proinsulin into insulin is a slow process which takes 30 to 120 minutes, beginning at the time of granule formation and continuing inside the newly formed granules. Conversion is thought to involve the sequential action of specific proteases, with trypsin-like and carboxypeptidase-B activities, present in the storage granules. Cleavage of the proinsulin molecule produces insulin and C-peptide and releases the two pairs of basic amino acids from each end of the connecting peptide (Figure 3.1). Removal of the connecting peptide decreases the solubility of the insulin molecule and in the presence of zinc, which is concentrated in the storage granule, insulin begins to precipitate as microcrystals of zinc-insulin hexamers. The precipitation of insulin in this crystalline state allows it to be stored in the most concentrated form and renders the molecule resistant to further enzymic degradation by the converting enzymes.

The conversion of proinsulin into insulin is accompanied by changes in the morphology of the granules from the immature granules with a low uniform density, to the mature granules with a dense central inclusion of crystalline zinc-insulin. The C-peptide probably remains in the clear space surrounding the dense central core as there is no evidence that C-peptide cocrystallises with insulin. Enzymic conversion of proinsulin into insulin proceeds slowly as the immature granules become mature storage granules and move from the area of the Golgi into the general cytoplasm of the B-cell to await the activation of the secretory process.

Proinsulin

Human proinsulin is a single chain polypeptide, composed of 86 amino acids and with a molecular weight of approximately 9,000 daltons. It contains the A and B chains of the insulin molecule in the same conformation as they occur in insulin, but they are linked by a connecting peptide of 35 amino acids which joins the *C*-terminus of the B chain to the *N*-terminus of the A chain (Figure 3.1). The connecting peptide is folded over a portion of the surface of the insulin moiety largely masking the hormone's active site and enabling proinsulin to exhibit only

5 per cent of the biological activity of insulin (Figure 3.5). It does not, however, obscure those surfaces which interact to form dimers and hexamers, or the sites of interaction with insulin antibodies.

The structures of the proinsulin connecting peptides from a variety of species, including man, have been determined. The connecting peptides vary in length from 26 (canine) to 35 (human) amino acids and their primary structures differ to a much greater extent than the primary structures of the corresponding insulin molecules. However, they all have a pair of basic amino acids at each end of the connecting peptide with two arginine residues at the *N*-terminus and a lysine and arginine at the *C*-terminus. These basic residues are the sites at which proteolytic

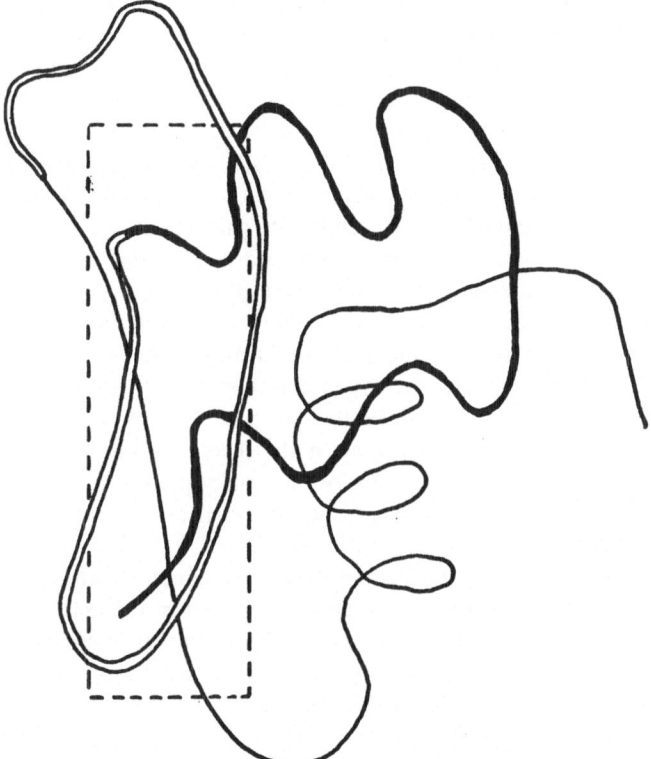

Figure 3.5: Probable conformation of human proinsulin in solution. ——— = B-chain; ——— = A-chain; ===== = Connecting peptide. The connecting peptide lies over the region of the molecule containing the active centre and effectively blocks the biological activity of the molecule.

cleavage of the molecule occurs during its conversion into insulin and being exposed on the surface of the molecule they are readily accessible to enzymic attack. In addition, certain regions within the connecting peptide segment are conserved and these may have important directional influences on the folding of proinsulin during biosynthesis.

The major function of the connecting peptide is to ensure that the proinsulin molecule folds in such a way as to achieve the correct alignment of the three disulphide bridges. That this function lies within the primary structure of the proinsulin molecule is clearly shown by the ability of reduced proinsulin to be readily reoxidised to its original form in high yield (70-80 per cent). It is not known why the connecting peptide contains many more than the three amino acids that are required to bridge the short gap between the N-terminus of the A chain and the C-terminus of the B chain. A number of possible reasons for this apparently excessive length have been proposed. It could help rapid and efficient enzymic cleavage by ensuring the correct orientation of the pairs of basic amino acids located at the sites of cleavage. It may also be important in increasing the length of the prohormone to enable it to span the membrane ribosome complex to gain access to the endoplasmic reticulum. Another important function may be related to the observation that the connecting peptide covers the surface of the proinsulin molecule masking a hydrophobic region. This would increase the solubility and prevent the precipitation and granulation of the proinsulin during its passage from the rough endoplasmic reticulum to the Golgi. The conversion of proinsulin into insulin in the B-granules is not fully completed and about 5 per cent of the secretory output of the B-cell is proinsulin. Proinsulin is therefore normally present in the circulation and the mean fasting concentration in normal subjects is in the region of 0.2 ng/ml.

It is not yet possible to measure proinsulin by direct immunoassay of unextracted human serum, as the antibodies used in the assay also react with insulin and C-peptide, which are both present in serum. A preliminary gel-filtration step is therefore required to separate proinsulin from insulin and C-peptide. Proinsulin normally constitutes about 15 per cent of the total insulin immunoreactivity of peripheral serum. This is a much higher percentage than that found in the portal vein or in the pancreas. This discrepancy is because proinsulin has a much longer half-life (20-30 min) in the circulation than insulin (4-5 min). The long half-life of proinsulin is related to the fact that, unlike insulin, it is only slowly degraded by the liver. Most of the proinsulin is removed from the circulation by the kidney.

Since proinsulin has a low biological activity (5-10 per cent of that of insulin) and is not converted into insulin by peripheral tissues, it has been suggested that an altered proinsulin/insulin ratio might occur in patients with diabetes. An increase in the proinsulin/insulin ratio has been reported in some cases of type 2 diabetes, although the significance of this finding is unclear.

Markedly elevated fasting serum concentrations of a proinsulin-like material with normal insulin concentrations have been reported in cases of familial hyperproinsulinaemia. This condition is inherited as an autosomal dominant trait and results from a mutation in the structural gene for insulin which affects the sites of cleavage of proinsulin to insulin. An amino acid substitution at these sites blocks the conversion of proinsulin into insulin and the patients have high circulating levels of intermediates of proinsulin conversion.

The major clinical significance of a raised serum proinsulin concentration has been in the diagnosis of islet cell tumours. Patients with islet cell tumours may show normal basal insulin concentrations but the proinsulin is invariably increased. In addition, it has been shown that the more malignant the tumour the higher is the circulating proinsulin concentration.

C-peptide

C-peptide is that region of the connecting peptide which remains intact after the proteolytic cleavage of proinsulin to insulin. In man it consists of the 31 amino acids remaining after the removal of the two basic residues from each end of the connecting peptide (Figure 3.1).

The sequestration of the proteolytic cleavage of proinsulin to insulin within the Golgi lamellae and immature B-granules results in the retention of equimolar amounts of C-peptide and insulin in the mature secretory granules. Thus, the B-cell secretory products consist of equimolar amounts of insulin and C-peptide (approximately 95 per cent) and small amounts of proinsulin and intermediate cleavage forms (approximately 5 per cent). The secretion of C-peptide along with insulin appears to be merely a convenient method of disposal of a synthetic byproduct since it has no detectable biological activity. The amino acid sequences of C-peptides from several species have been determined. These peptides exhibit a much higher rate of mutation acceptance than do the corresponding insulins, a finding which is consistent with the lack of biological activity of the C-peptide.

The normal fasting concentration of C-peptide in the peripheral blood of man is in the region of 0.9-3.5 ng/ml. This is more than five times higher, on a molecular basis, than that of insulin and is due to its prolonged half-life in the circulation. The prolonged half-life of C-peptide is related to the fact that unlike insulin it is not rapidly metabolised by the liver and it is cleared from the circulation largely by the kidney.

The immunoassay of insulin has been widely used to study insulin levels in healthy subjects and diabetics managed with diets and oral hypoglycaemic agents, although the method has proved less useful in insulin-treated diabetic patients. The reasons for this include the production in some patients of circulating insulin antibodies in response to injected insulin, which interfere with the assay procedure. In addition, the immunological similarity of the injected insulin compared to the human hormone means that they cannot be distinguished by immunoassay. Immunoassay procedures specific for C-peptide are not affected by insulin nor by insulin antibodies, and since circulating C-peptide concentrations directly reflect B-cell secretory activity, the estimation of the serum C-peptide has provided a means of studying B-cell secretory capacity in these patients. Measurement of serum C-peptide in patients with established diabetes has shown varying degrees of B-cell secretory impairment. In general, it has been found that retention of B-cell secretory ability in diabetic patients, albeit at a lower than normal level, may facilitate good control. In those patients without significant B-cell reserve, meticulous attention to diet and exogenous insulin therapy is necessary for good control. These studies of C-peptide levels have also shown that loss of B-cell secretory capacity in type 1 diabetes may not be abrupt, but can continue for several years after the diabetes becomes clinically manifest.

One of the most useful applications of the C-peptide assay has been to facilitate the diagnosis of various hypoglycaemic disorders, including islet cell tumors and suspicious hypoglycaemia appearing in diabetics and non-diabetic subjects. Using the assay it is possible to determine whether the hypoglycaemia is related to abnormal endogenous insulin secretion or the administration of exogenous insulin since in the former there is an associated increase in C-peptide levels, while in the latter there is no such increase.

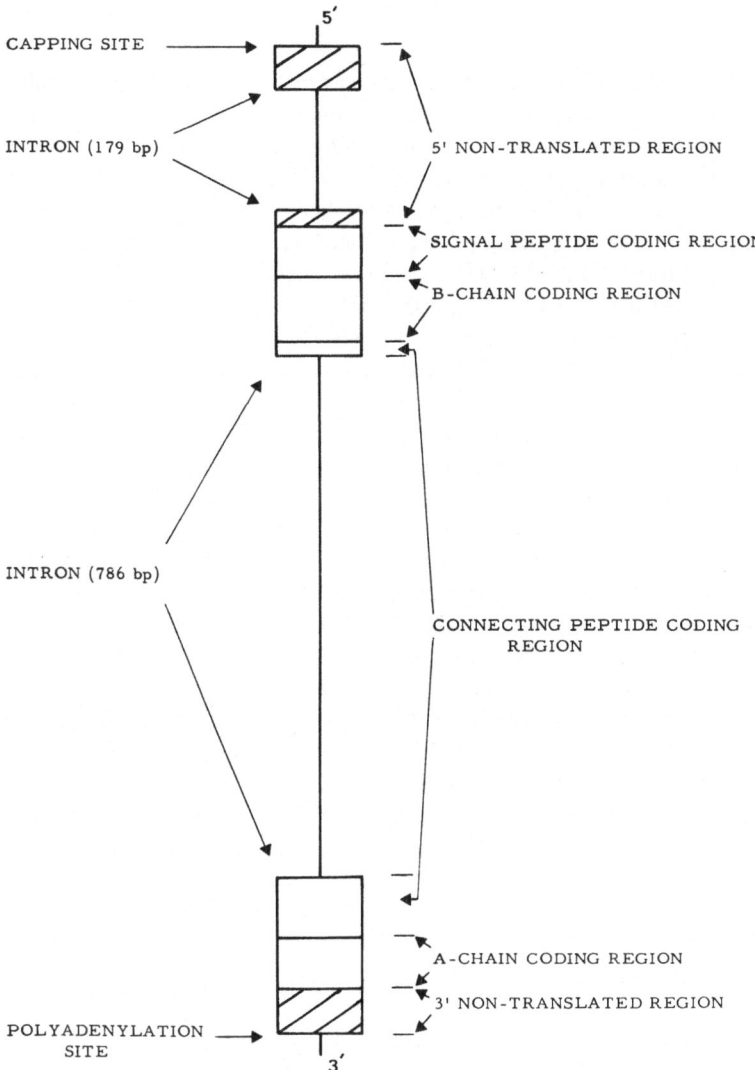

Figure 3.6: Diagrammatic Representation of the Human Insulin Structural Gene

Insulin Gene

Recent advances in molecular and cellular biology have made it possible to determine the structure of the human insulin gene. It appears that the human haploid genome contains a single insulin gene since only one insulin protein has been isolated. Somatic cell hybrids formed between cultured mouse and human cells have been used to map the human genome and these studies have localised the insulin gene to the short arm of chromosome 11. The structure of the gene has recently been determined using recombinant DNA technology (Figure 3.6). The gene encodes a 1430-nucleotide insulin mRNA precursor that contains two intervening sequences of 179 and 786 nucleotides. The first of these sequences is located in the 5′-untranslated region of the mRNA and the other is located in the C-peptide encoding region. These intervening sequences are excised from the messenger RNA precursor as it is processed. The product of this processing is the insulin mRNA molecule which directs the synthesis of preproinsulin (Figure 3.7).

A highly polymorphic region of DNA near the human insulin gene has been detected by restriction endonuclease mapping of DNA fragments from a variety of individuals. This polymorphism results from DNA insertions or deletions located approximately 500 nucleotides from the 5′ end of the insulin gene. The insertions in this region are primarily of two size classes, 0-600 nucleotides and 1,600-2,200 nucleotides. The physiological significance of this polymorphism is not known. However, the demonstration that type 2 diabetics have 6.3-fold increase in the frequency of the larger insertion compared to type 1 diabetics suggests that it may provide a genetic marker for type 2 diabetes.

Insulin Secretion

Glucose homeostasis depends on a variety of hormonal responses to increase the blood glucose concentration, but relies almost entirely on just one hormone, insulin, to lower the concentration. Alterations in plasma insulin levels and hence its activity are regulated largely by changes in the rate of insulin secretion and there is much evidence to suggest that one of the major factors which produces carbohydrate intolerance and leads to the development of diabetes mellitus is a derangement of the secretory activity of the B-cell. Much effort has therefore gone into trying to understand the insulin secretory process in

INSULIN STRUCTURAL GENE

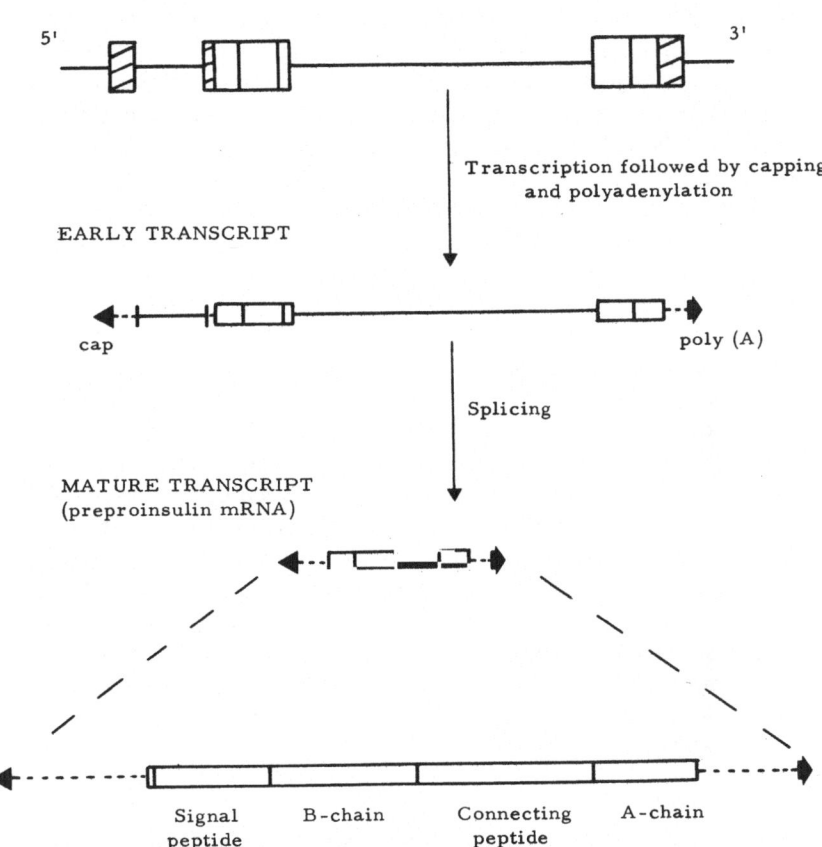

Figure 3.7: Transcription of the Human Insulin Structural Gene

molecular terms. It is to be hoped that such a detailed understanding will provide an insight into the derangements responsible for diabetes and will enable therapeutic measures to be developed which correct the derangements. The insulin secretory process is the most extensively studied of all the peptide hormone secretory processes and it would clearly be impossible to cover all aspects of this fascinating subject. In this section I have attempted to pick out those features which appear to provide the basis for further studies aimed at defining in detail the molecular basis of this process. There are many good reviews on different aspects of this subject and some of them are listed at the end of this chapter. They should be consulted if further information is required. The development of radioimmunoassay procedures for the assay of insulin in nanogram quantities and the introduction of techniques for the isolation of functionally intact islets of Langerhans from the pancreas have both been major factors in making many of these studies possible.

Studies on the insulin secretory process and its regulation have revealed a number of features which are common to all tissues which store their secretory products in membrane-limited vesicles prior to release, and, in addition, a number of features have emerged which appear to be unique to the B-cell. Of particular interest in the latter context is the wide range of physiological agents which affect insulin secretion (Table 3.3). The regulation of a single process by such a variety of factors implies that the B-cell is equipped with a number of distinct sensor systems able to identify and quantify each regulatory factor. In addition, the coupling between such sensor systems and the effector system responsible for the margination and exocytosis of the insulin secretory granules, is likely to involve a variety of signal molecules. Two such signal molecules have been identified in the B-cell, calcium and cyclic AMP, and they both appear to play an important role in stimulus-secretion coupling. Calcium appears to be the primary signal responsible for activating the secretory process and cyclic AMP appears to modulate the size of the secretory response.

General Features of the Secretory Response of the B-cell to Physiological Agents

The secretory response of the B-cell is controlled on a minute to minute basis by fluctuations in the extracellular concentrations of several hormones and neurotransmitters and a variety of nutrients such as glucose, amino acids, fatty acids and ketone bodies. Certain of the hormones and all of the nutrients reach the B-cell via the islet vascular system, while other hormones and the neurotransmitters are released locally from other endocrine cells or from autonomic nerve fibres ending in the islet. In addition, there are long-term adaptive changes in the secretory activity of the B-cell in response to changes in the physiological status of the animal. These long-term adaptive changes, like the short-term alterations, appear to be regulated by nutrient and hormonal factors.

The physiological agents which affect insulin release can be classified as primary stimuli, agents which promote insulin release directly;

secondary stimuli, agents which alter the response to a primary stimulus but do not directly affect the secretory rate themselves; and inhibitors, agents which inhibit the response to both primary and secondary stimuli (Table 3.3).

Table 3.3: Agents Which Affect Insulin Secretion

Primary Stimuli	Secondary Stimuli	Inhibitors
(a) Physiological		
Glucose	Glucagon	Somatostatin
Mannose	Secretin	Adrenaline
Leucine	Pancreozymin	Noradrenaline
Arginine	Gastrin	
Lysine	Acetylcholine	
Short-chain fatty acids	Prostaglandin E_1 and E_2	
Long-chain fatty acids		
Acetoacetate		
β-Hydroxybutyrate		
(b) Pharmacological		
N-Acetylglucosamine	Theophylline	Diazoxide
Glyceraldehyde	Caffeine	Mannoheptulose
Dihydroxyacetone	Isobutyl-methylxanthine	2-Deoxyglucose
Glucosamine	Sulphonylureas	Iodoacetate
Inosine		

This fundamental difference between the two classes of stimuli suggests a difference in their mode of action on the release process, and may reflect the fact that the interaction of a primary stimulus with its sensor mechanism generally results in the production of a calcium signal while the interaction of a secondary stimulus with its sensor system often results in the production of a cyclic AMP signal.

The response of the B-cell to all physiological stimuli is rapid, occurring within seconds of the addition of the stimulus and there is a rapid cessation of the response when the stimulus is removed. In addition, the response depends on the presence of functional microtubular and microfilamentous sytems within the cell, on the maintenance of intracellular ATP and cyclic AMP concentrations, and on the presence of extracellular calcium.

Calcium and the Regulation of Insulin Secretion

The secretion of insulin from the B-cell in response to all physiological stimuli requires the presence of extracellular calcium. This requirement may simply indicate that calcium is a necessary cofactor for the functioning of the secretory mechanism although numerous observations suggest a more direct role for calcium in secretion. Thus, an increase in the concentration of ionised calcium within the cytoplasmic compartment of the B-cell, induced by stimuli, is thought to trigger the secretory mechanism. In this way, calcium is thought to act as an important signal in the B-cell, linking stimulus recognition events to the secretory process.

The calcium concentration in the cytoplasm of the B-cell is determined by the relative rates of its influx and efflux across the plasma membrane and by the relative rates of uptake and release from other intracellular compartments, including mitochondria, endoplasmic reticulum and storage granules. The entry of calcium into the B-cell is thought to involve a number of different routes including a specific calcium channel which is sensitive to membrane potential.

In the non-stimulated B-cell the cytoplasmic calcium concentration is maintained in the region of 10^{-7} M, whereas the extracellular calcium concentration is normally 10^{-3} M. Thus, there is a concentration gradient favouring the entry of calcium into the B-cell. This is augmented by an electrical gradient since the cell interior is electronegative relative to the exterior. This electrochemical gradient favours calcium entry into the B-cell, but in the unstimulated cell the rate of influx is usually low since the membrane is relatively impermeable to the cation. The resting membrane potential of the B-cell is around -50 to -70 mV and its maintenance depends mainly on K^+ permeability. In response to a primary stimulus, such as glucose, there is a rapid depolarisation of the plasma membrane as a result of a reduction in its K^+ permeability. This depolarisation opens the voltage-dependent calcium channel in the membrane and there is a rapid accumulation of calcium within the cytoplasm of the B-cell. When the calcium concentration reaches a critical value the release mechanism is triggered.

Secondary stimuli, unlike primary stimuli, do not appear to alter the net uptake of calcium into the B-cell. However, these agents may affect the intracellular distribution of the cation within the cell by promoting the efflux of calcium from intracellular storage pools. Thus, secondary stimuli may increase the response of the B-cell to a primary stimulus by increasing the availability of calcium in the cytoplasmic compartment. In the absence of a primary stimulus, such increased availability of calcium does not appear to have an effect on the release mechanism, possibly because the calcium is rapidly transported across the plasma membrane. This transport of calcium is thought to be linked to calcium-stimulated ATPase activity and sodium/calcium exchange mechanisms in the plasma membrane which pump calcium from the cell against a concentration gradient. The major intracellular storage pools of calcium in the B-cell appear to be confined within the inner mitochondrial membrane and the endoplasmic reticulum. The release of calcium from the mitochondrial pool is stimulated by cyclic AMP and the intracellular concentration of cyclic AMP may therefore play an important role in regulating the subcellular distribution of calcium in the B-cell.

Cyclic AMP and the Regulation of Insulin Secretion

Unlike the pancreatic B-cell, most secretory tissues respond to only a limited number of physiological stimuli. In many instances, the effect of the stimulus on the release process appears to be mediated via the cyclic AMP system. Interaction of the stimulus with specific receptor units on the plasma membrane activates adenylate cyclase and increases the intracellular concentration of cyclic AMP. This, in turn, activates the release mechanism via an effect on cyclic AMP-dependent protein kinase activity (Figures 4.2 and 5.3). The response in such a system is rapid, specific and quantitatively related to the magnitude of the stimulus. A regulatory system of this type appears to operate in a variety of secretory tissues, including the thyroid and anterior pituitary and there is considerable evidence to suggest that the effects of secondary stimuli on insulin secretion are mediated via the cyclic AMP system of the B-cell.

Thus, the pancreatic B-cell contains all the components of the cyclic AMP system, including the plasma membrane adenylate cyclase, soluble and particulate cyclic nucleotide phosphodiesterase activities and soluble and particulate cyclic AMP dependent protein kinase activities. Furthermore, changes in the intracellular cyclic AMP concentration of the B-cell follow its interaction with a number of secondary stimuli. These changes in cyclic AMP are the result of alterations of adenylate cyclase or cyclic nucleotide phosphodiesterase activities caused directly by the secondary stimulus. Since secondary stimuli do not stimulate secretion on their own it is likely that cyclic AMP is not able to trigger the release mechanism but it is clearly able to modulate the size of the secretory response.

Primary stimuli may also increase the intracellular cyclic AMP concentration in the B-cell but this effect appears to be secondary to the increase in cytoplasmic calcium induced by these agents. This is because the major receptor for calcium in the cytoplasm of the B-cell is the calcium-binding protein calmodulin and the interaction of calcium with this protein produces a calcium-calmodulin complex which has many actions (Figure 3.8) including the activation of the B-cell adenylate cyclase.

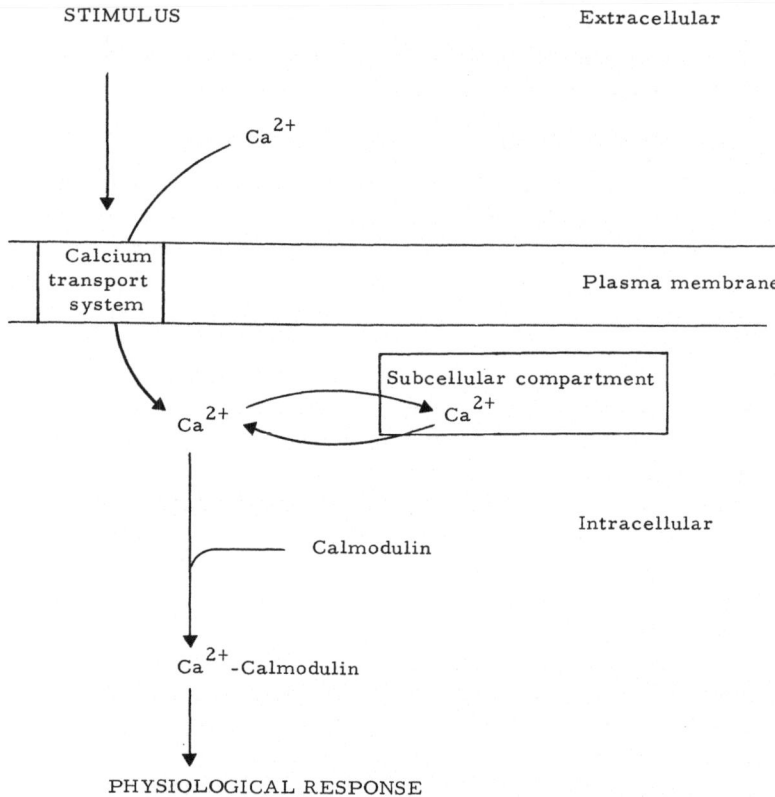

Figure 3.8: The Calcium-Calmodulin System

Glucose and the Regulation of Insulin Secretion

An increase in the blood glucose concentrations above the fasting level of 4-5mM is the major physiological stimulus to insulin release. Since there are alterations in the response of the B-cell to this stimulus in diabetes mellitus, it is important to determine how the B-cell recognises an increase in extracellular glucose concentration and how this recog-

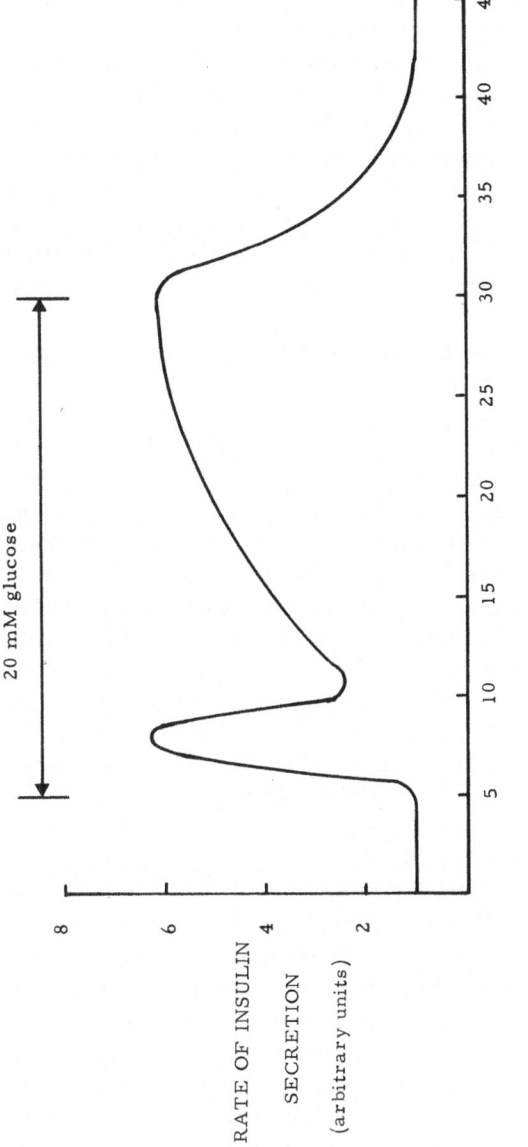

Figure 3.9: Dynamics of the Insulin Secretory Response to Glucose

nition is translated into an elevation in the cytoplasmic concentration of ionised calcium to activate the release mechanism.

The response of the B-cell to glucose is rapid and there is a biphasic pattern of release with a spike-like first phase followed by a nadir and a slowly rising second phase (Figure 3.9). The molecular basis of this biphasic pattern of secretion is not known although it may reflect the presence of two pools of secretory granules in the B-cell. During the first phase those granules that are near the plasma membrane are released, there is then a delay while granules are moved to the surface via the microtubular system. The second sustained phase involves these newly mobilised granules.

The secretory response to glucose is relatively specific since galactose and fructose are both without effect. In addition, there is a sigmoid relationship between the secretory response and the extracellular glucose concentration (Figure 3.10). The most dramatic effects on secretion are observed at glucose concentrations just above the fasting level. The sensitivity of the B-cell to this range of glucose concentrations relates to the function of the hormone in the maintenance of blood glucose concentrations at the fasting level. The B-cell must therefore be equipped with a specific glucose recognition unit which both senses and responds to changes in the extracellular glucose concentration. Such a recognition unit need not be located on the plasma membrane since there is a rapid equilibration of intracellular and extracellular glucose concentrations in the B-cell.

The glucose recognition unit is likely to be a protein which is capable of binding glucose with low affinity ($\sim 8 \times 10^{-3}$ M) but high specificity. The low affinity of the receptor is necessary to ensure that the extent of binding responds to changes in the physiological concentration of glucose. Furthermore, the binding of glucose to the recognition unit should initiate the events that lead to membrane depolarisation and calcium entry. It seems likely that one or more of the rate-limiting enzymes of glucose metabolism in the B-cell may act as the glucose recognition unit. The activity of these enzymes is thought to be sensitive to glucose concentration and they therefore regulate glucose metabolism in response to glucose concentration. In this way they may generate one or more metabolic intermediates or cofactors which are responsible for membrane depolarisation. Evidence in favour of this hypothesis comes from the observations that glucose has to be metabolised in the B-cell to promote insulin release. Thus, increased metabolism of glucose in the B-cell is paralleled by increasing rates of insulin secretion; sugars, or their derivatives which are poorly metabolised, are weak or inefficient

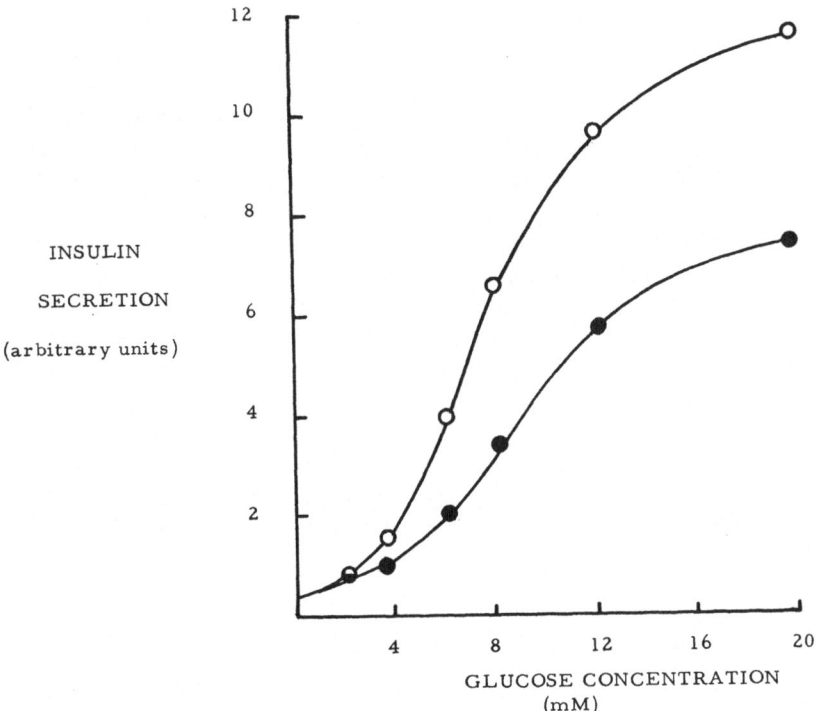

Figure 3.10: Effect of Glucose Concentration on Insulin Secretion in the Presence and Absence of a Secondary Stimulus. ● = Glucose alone; ○ = Glucose + secondary stimulus.

stimulants; metabolic inhibitors such as mannoheptulose, 2-deoxyglucose and iodoacetate reduce glucose-stimulated secretion and the triose glyceraldehyde which is metabolised via glycolysis mimics the secretory efficacy of glucose. This requirement for glucose metabolism is in part related to the generation of ATP which is required by the secretory mechanism. In addition, however, glucose metabolism is thought to produce an intermediate or cofactor which is responsible for the depolarisation of the plasma membrane and the consequent increase in its permeability to calcium. Recent evidence suggests that it is a change in the NAD(P)H/NAD(P) ratio in the cytoplasm of the B-cell generated via the metabolism of glucose through glycolysis that is responsible either directly or indirectly for the membrane depolarisation.

The rate of flux of glucose through glycolysis is thought to be an important determinant of intracellular NAD(P)H/NAD(P) ratios. Accordingly, the glucose recognition unit may be one or more of the rate-limiting enzymes of glycolysis. The initial phosphorylation of glucose is the first rate-limiting step of glucose metabolism in the B-cell since the rate of transport of glucose into the cell is much faster than the rate of glucose metabolism. The B-cell is equipped with a low affinity ($\sim 10 \times 10^{-3}$ M), high capacity glucose phosphorylating enzyme (glucokinase) whose activity is clearly regulated by physiological concentrations of glucose and this enzyme may well be the glucose recognition unit. In addition, the phosphofructokinase step is also an important regulator of glycolysis. This is the first enzyme unique to glycolysis and its activity is known to be modulated by a number of allosteric factors. It is possible therefore that glucose may directly or indirectly affect the activity of this enzyme and thereby regulate glycolytic flux in the B-cell. If this were so, then it would mean that phosphofructokinase is the glucose recognition unit. In this case the interaction of glucose with the recognition unit would be that of an allosteric interaction whereas in the case of glucokinase the interaction would be that of a substrate interacting with the catalytic site of an enzyme.

Short-Term Regulation of Insulin Secretion

The major factor responsible for the short-term regulation of insulin secretion is the concentration of various nutrients in the circulation (Table 3.3). The most important of these is glucose, although amino acids, fatty acids and ketone bodies may also stimulate secretion under certain circumstances.

The responsiveness of the pancreatic B-cell to the circulating concentration of nutrients can be modified, in the short term, by changes in the extracellular concentration of a variety of hormones. Thus, glucagon, pancreozymin and secretin all potentiate (Figure 3.10) the response of the B-cell to glucose, an effect which is mediated by the cyclic AMP system of the B-cell. Pancreozymin and secretin concentrations are generally increased following the ingestion of a meal and the potentiation of insulin secretion observed with these hormones may therefore be important in achieving a circulating insulin concentration which is adequte to ensure the rapid removal of metabolites from the circulation following their absorption.

The catecholamines, adrenaline and noradrenaline inhibit the release of insulin in response to all metabolite stimuli. This effect is related to their ability to activate *a*-receptors on the B-cell, since phentolamine, an α-receptor blocker, abolishes the inhibitory effect, while β-receptor blockade with propranolol is without effect. Activation of α-receptors in the B-cell inhibits adenylate cyclase, lowers intracellular cyclic AMP levels and may also affect calcium handling. As a consequence the secretory response to any stimulus is reduced. The inhibitory effects of catecholamines on insulin release appear to be physiologically important in maintaining elevated blood glucose concentrations arising from catecholamine-induced glycogenolysis during stress and exercise.

In addition to the hormones which reach the B-cell via the circulation, the B-cell is likely to be affected by hormones released locally from the other cells of the islet. These include glucagon, somatostatin and pancreatic polypeptide released from the endocrine cells, prostaglandins released from the endothelial cells of the capillaries which permeate the islets and acetylcholine and noradrenaline released from nerve endings in contact with the islet cells.

Whilst hormonal and nutrient factors provide the major controls of insulin secretion it is clear that both major branches of the autonomic nervous system also contribute to the control of insulin secretion. Nerves to the islets contain sympathetic and parasympathetic fibres and both adrenergic and cholinergic terminals are found in the islets.

The sympathetic nervous system may affect insulin secretion either by the release of catecholamines from the adrenal medulla or by the release of neurotransmitters from nerve terminals in the islets. Both adrenaline, the major adrenomedullary hormone, and noradrenaline, the adrenergic neurotransmitter, inhibit insulin secretion. These inhibitory effects of the sympathetic nervous system on insulin release may be important during stress when there is a need to elevate blood glucose concentration. For instance, the suppression of insulin release that accompanies a fall in systemic blood pressure is dependent on sinus nerve baroreceptor afferents and splanchnic nerve connections to the islets.

Acetylcholine, a parasympathetic neurotransmitter, stimulates insulin secretion and this may be important following the ingestion of a meal. Thus, the early rise in insulin release after oral ingestion appears to involve gustatory stimulation relayed to the islets via the vagus.

There are a number of pharmacological agents which stimulate insulin release (Table 3.3). In general, they appear to act as secondary stimuli and to affect the secretory mechanism via the cyclic AMP system. The most important clinically of these agents are the sulphonylureas since these are widely used to stimulate insulin release in type 2 diabetics (Chapter 9).

Long-Term Regulation of Insulin Secretion

In addition to the acute, minute-by-minute response of the B-cell to fluctuations in the extracellular concentration of nutrients and hormones, there exist in the B-cell regulatory mechanisms responsible for longer-term adaptations of the secretory response. Such mechanisms operate to maintain an enhanced responsiveness of the B-cell to metabolic stimuli during pregnancy, and a decreased responsiveness during starvation. The intracellular regulatory system which is involved in these adaptations appears to be the cyclic AMP system of the B-cell since B-cell cyclic AMP concentrations and adenylate cyclase activity are increased during pregnancy and decreased during starvation.

Mode of Action of Calcium and Cyclic AMP on the Insulin Secretory Mechanism

The intracellular concentrations of calcium and cyclic AMP in the B-cell play important roles in linking the events related to stimulus perception to the activation of the secretory mechanism, and both are necessary to ensure the correct magnitude of response. Calcium appears to mediate the effects of primary stimuli on the release process while cyclic AMP mediates the effects of secondary stimuli. However, there are complex interrelationships between the two signal molecules since a rise in cytosolic calcium concentration can increase cyclic AMP levels, while a rise in cyclic AMP concentration can increase cytosolic calcium levels. It seems likely therefore that these two signal molecules work in a concerted manner to ensure that the magnitude of the secretory response is entirely appropriate to the prevailing extracellular environment.

The action of cyclic AMP on the secretory mechanism appears to involve the activation of cyclic AMP-dependent protein kinase enzymes which phosphorylate and thereby alter the functional activity of components directly involved in the secretory mechanism. In addition, alterations in the extent of phosphorylation of proteins in the inner mitochondrial membrane may promote the efflux of calcium from the mitochondria.

The mode of action of calcium on the secretory mechanism is unknown at present, although it is likely to involve the direct interaction of calcium with functional components of the release mechanism. In addition, calcium may also regulate the activity of protein kinase enzymes in the B-cell. While calcium may interact directly with certain protein kinase enzymes, the activation of most protein kinases by calcium involves the calcium binding protein calmodulin. Calmodulin is present in the B-cell and in the presence of calcium a calcium-calmodulin

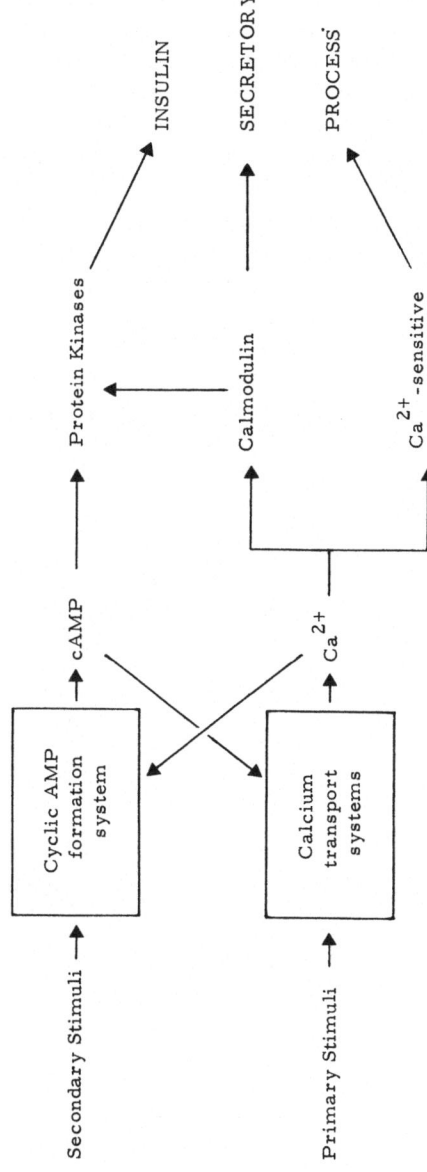

Figure 3.11: Regulation of Insulin Secretion

complex is formed which activates the protein kinase. Thus the degree of phosphorylation of protein components of the secretory mechanism, regulated by the intracellular concentrations of both calcium and cyclic AMP, may control the activity of secretory process (Figure 3.11).

Insulin released during the first hour of stimulation comes exclusively from the cytoplasmic pool of insulin-storage granules. Under conditions of maximal stimulation by glucose, less than 10 per cent of the insulin stored in the B-cell is released and since glucose also has a dramatic effect in promoting insulin biosynthesis there is normally always a considerable reserve of insulin in the B-cell. The release mechanism has two major components, the movement of the insulin-storage granules to the plasma membrane (margination) and the extrusion of the granule contents through an opening created by the fusion of the plasma membrane with the granule membrane (exocytosis). These events are illustrated in Figure 3.12 and they are all possible sites for the action of cyclic AMP and/or calcium in regulating secretion.

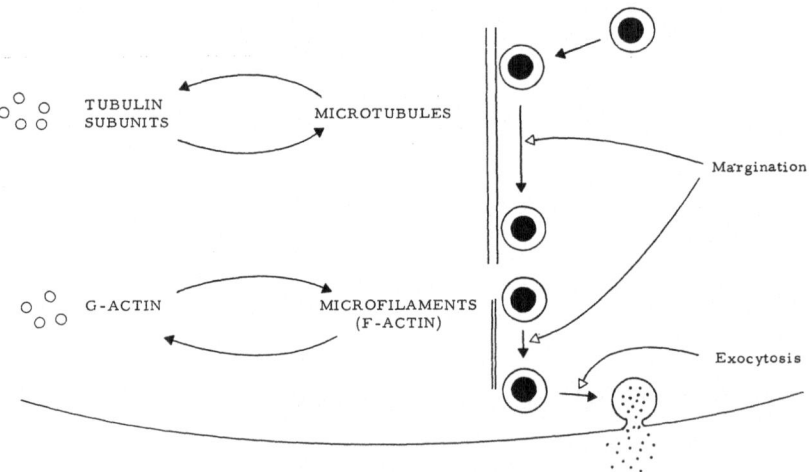

Figure 3.12: Ultrastructural Aspects of the Insulin Secretory Process

Margination

In the unstimulated B-cell, the insulin-storage granules are randomly distributed throughout the cytoplasm. However, during stimulation there is a relative increase in the number of granules close to the cell surface due to the directed movement of granules towards the plasma membrane. Intracellular movement in a variety of cell types is thought to be associated with the network of microtubules and microfilaments observed in the cell. It is thought that the microtubular system transports the granules over relatively long distances, from the general cytoplasmic storage pool towards the plasma membrane, while the microfilamentous system is responsible for the final delivery of the granule to the membrane.

Evidence that the microtubular system of the B-cell might play an important role in secretion, was provided by the observation that agents such as colchicine and vinblastine, which disrupt the microtubular system, inhibit the secretion of insulin. It was originally suggested that the microtubular network might provide a contractile system for granule movement in the B-cell, although it now seems more likely that they provide intracellular pathways guiding the movement of granules. Calcium may play a role in regulating the interaction between storage granules and microtubules, an interaction which might be necessary to ensure the vectorial transport of granules in the B-cell.

Microtubules are polymerised structures in a state of dynamic equilibrium with a pool of subunits (tubulin). Tubulin is a protein of molecular weight composed of two different monomeric units of molecular weight 56,000 (a-tubulin) and 50,000 (β-tubulin). In the unstimulated B-cell about 30 per cent of the microtubular protein is present as polymerised microtubules, the remaining 70 per cent being free subunits. When insulin secretion is stimulated by glucose or by an increase in intracellular cyclic AMP, there is a rapid shift in the equilibrium in favour of polymerised microtubules. Under conditons of maximal stimulation over 70 per cent of the microtubular protein is present as polymerised microtubules. Both calcium and cyclic AMP are thought to control the relative rates of polymerisation and depolymerisation of microtubules and this may be an important site for their action in regulating insulin secretion. The effect of cyclic AMP in promoting polymerisaton appears to be mediated via the activation of a cyclic AMP-dependent protein kinase since the phosphorylation of microtubular protein or proteins closely associated with microtubules has been shown to occur in response to cyclic AMP.

Evidence that the microfilamentous system of the B-cell plays an important role in regulating insulin secretion was provided by the observation that cytochalasin B, an agent which disrupts the microfilamentous system, has dramatic effects in promoting insulin release. This observation led to the suggestion that the cell web, under non-stimulatory conditions, acted to prevent the storage granules reaching the plasma membrane and that under stimulating conditions, contraction of the elements of the system brought the granules in contact with the plasma membrane. The presence of the microfilament protein, actin, as a component of the cell web, is compatible with this role for the cell web. In addition, it is likely that calcium may play a role in regulating the contractile activity of the cell web since calcium controls the polymerisation and activity of other actin-containing contractile systems.

Microfilaments are thought to be composed largely of F-actin (filamentous actin) which is produced by the polymerisation of G-actin (globular actin). G-actin is a single polypeptide chain globular protein of molecular weight 46,000 daltons which readily polymerises to helical filaments of F-actin. It has been shown that during active secretion there is an increased conversion of G-actin into F-actin, an event which may lead to an increase in the rate of access of the storage granules to the plasma membrane, since storage granules have been shown to interact with F-actin but not G-actin. Calcium may play a role in promoting the interaction of actin monomers to form F-actin and this action may involve additional protein factors.

Myosin has been identified as a major cellular protein in islet cells and it appears to resemble the protein in smooth muscle since it is composed of a heavy chain (200,000 daltons) and two light chains (14,000 and 19,000 daltons). The interaction of actin and myosin to form the actomyosin complex is an important component of many contractile systems and this complex may be involved in granule movement in the B-cell. The interaction of actin and myosin and hence the contractile activity of the complex depends on the phosphorylation of the light chains of myosin. This phosphorylation is carried out by myosin light chain kinase, a calcium/calmodulin dependent protein kinase. Thus, calcium may regulate granule movement in the B-cell by controlling the contractile activity of actomyosin complexes.

It is likely therefore that the movement of B-cell storage granules to the cell membrane involves the coordinated activity of the microtubular and microfilamentous systems of the B-cell and possibly a contractile system based on actomyosin. These systems are based on components formed by the polymerisation or interaction of protein subunits and the regulation of the polymerisation/depolymerisation rate by calcium and/or cyclic AMP may be important in controlling insulin secretion. In addition, the functional activities of these systems may involve reversible phosphorylation catalysed by cyclic AMP and/or calcium-dependent protein kinases. Furthermore, the interaction of the storage granules with the systems may also be regulated by calcium and/or cyclic AMP. Thus, there are a number of sites at which both cyclic AMP and calcium may act to control the margination of granules during secretion. These are shown diagramatically in Figure 3.12.

Exocytosis

The final event in the insulin secretory process involves the fusion of the membrane of the insulin storage granule with the plasma membrane of the B-cell, the granule contents being released from the cell at the site of membrane fusion (Figure 2.7). Once outside the cell, the crystalline contents rapidly dissolve in the extracellular fluid and they subsequently enter the circulation. Immediately prior to the fusion of the granule and plasma membranes, proteins are removed from the regions of the plasma membrane directly involved in the exocytotic event. This movement of protein may be associated with an increase in membrane fluidity arising from the increase in the degree of unsaturation of fatty acids which has been observed in membrane phospholipids during insulin secretion.

The membrane of the insulin storage granule and the plasma membrane of the B-cell must undergo biochemical modification before they can fuse, since cell membranes are inherently non-fusagenic. The major barrier to membrane fusion appears to be the highly charged head groups on the membrane phospholipids. It has been suggested that calcium may promote membrane fusion either by neutralising these charges and allowing the membranes to come close together, or by serving as a bridge linking phospholipids in the two membranes. However, it seems more likely that the removal of the head groups with the conversion of membrane phospholipids into 1:2-diacylglycerols is the event which triggers membrane fusion since diacylglycerols promote membrane fusion in a variety of situations. The enzyme activity responsible for this alteration is phospholipase C and the activity of the enzyme is regulated by calcium. Thus, calcium may control, not only the rate of movement of granules to the plasma membrane, but also the final membrane fusion event.

In the process of exocytosis the granule membrane fuses with the plasma membrane and it has been estimated that during a period of sustained secretory activity lasting for 60 minutes an area of granule membrane equal to that of the plasma membrane fuses with the plasma membrane. Accordingly, the plasma membrane should enlarge to twice its original size. This, however, does not occur because areas of the plasma membrane are removed from the surface via the formation of endocytotic vesicles.

Further Reading

Ashcroft, S.J.H. Glucoreceptor Mechanisms and the Control of Insulin Release and Biosynthesis. *Diabetologia* (1980) *18*, 5-15

Bell, G.I. *et al*. Sequence of the Human Insulin Gene. *Nature* (1980) *284*, 26-32

Campbell, I.L. *et al*. Insulin Biosynthesis and its Regulation. *Clinical Science* (1982) *62*, 449-455

Chan, S.J. & Steiner, D.F. Preproinsulin, a New Precursor in Insulin Biosynthesis. *Trends in Biochemical Sciences* (1977) *2*, 254-256

Hedeskov, C.J. Mechanism of Glucose-Induced Insulin Secretion. *Physiological Reviews* (1980) *60*, 442-509

Kitabchi, A.E. Proinsulin and C. Peptide. *Metabolism* (1977) *26*, 547-587

Malaisse, W.J. *et al*. Insulin Release: Reconciliation of the Receptor and Metabolic Hypotheses. *Molecular & Cellular Biochemistry* (1981) *37*, 157-165

Montague, W. & Howell, S.L. Cyclic AMP and the Physiology of the Islets of Langerhans. Advances in *Cyclic Nucleotide Research* (1975) *6*, 201-242

Owerbach, D. *et al*. The Insulin Gene is Located on the Short Arm of Chromosome 11 in Humans. *Diabetes* (1981) *30*, 267-270

Rubenstein, A.H. *et al*. Clinical Significance of Circulating Proinsulin and C-Peptide. *Recent Progress in Hormone Research* (1977) *33*, 435-453

Steiner, D.F. Insulin Today. *Diabetes* (1977) *26*, 322-340

Tomlinson, S. *et al*. Calmodulin and Insulin Secretion. *Diabetologia* (1982) *22*, 1-5

Wolheim, C.B. & Sharp, G.W.G. Regulation of Insulin Release by Calcium. *Physiological Reviews* (1981) *61*, 914-973

4 INSULIN ACTION

Introduction

While there has been rapid progess in the study of the molecular mechanism of action of many hormones it is only recently that significant advances have been made in our understanding of the mechanism of action of insulin. These advances began in the early 1970s when it was established that the initial event in the action of insulin was its interaction with specific receptors on the plasma membrane of the target cell. This interaction has been defined in detail and recent studies have begun to reveal the structure of the receptor. The receptor serves to recognise the specific hormone in the plasma, to bind it and then, as the hormone-receptor complex, to activate the cell. It is, however, still unclear how the interaction of insulin with its receptor leads to the appropriate cellular response, although significant progress is being made in this direction.

Insulin Receptors

Insulin receptors have been demonstrated both on the classic target cells of insulin action (muscle, liver and adipose tissue) and on several other cell types such as lymphocytes, monocytes and granulocytes which appear to have no well-defined insulin response. The receptors on a variety of human tissues share many features in common with receptors on tissues from most other species. They all reversibly bind the hormone with high affinity and specificity. The high affinity $(\sim 10^{-10} \, M)$ of the insulin receptor for insulin is necessary to ensure that the receptor can bind insulin at its physiological concentration $(10^{-10} \text{-} 10^{-9} \, M)$. The biological specificity of the insulin receptor is one of its most characteristic features. Insulin and peptides with insulin-like activity compete for receptor binding in direct proportion to their biological activity, while hormones unrelated to insulin do not bind to the insulin receptor. Thus, chicken insulin, pig insulin, pig proinsulin and guinea-pig insulin bind with molar activities of 300:100:15:2

respectively, while their relative biological potencies are 300:100:16:1. The binding of insulin to its receptor is saturable, occurs rapidly and proceeds until a steady state is reached. Dissociation also occurs rapidly in response to a fall in insulin concentration and the insulin is released in an unmodified form. The total number of insulin receptors per cell varies widely in different human cells, but when expressed per unit area of cell surface the concentration of receptors shows less variation (Table 4.1).

Table 4.1: Insulin Receptors on Various Human Tissues

	Receptor Number (sites/cell)	Receptor Concentration (sites/μm^2 cell membrane)
Granulocytes	1×10^3	3
Monocytes	1.2×10^4	17
Lymphocytes	1.2×10^4	24
Adipocytes	3×10^5	15
Hepatocytes	3×10^5	65

Scatchard analysis of the kinetics of the insulin-receptor binding reaction has consistently produced a curvilinear plot. Initially, this led to the suggestion that there were two types of binding sites, a small number of high affinity ($\sim 10^{-10}$ M) sites and a large number of lower affinity ($\sim 10^{-9}$ M) sites. It appears more likely, however, that there is a single type of receptor and that the affinity of the receptor for insulin decreases with increasing levels of receptor saturation. This 'negative cooperativity' is thought to be due to a progressive cooperative inter-action between the receptors as they bind insulin which makes it increasingly more difficult to bind further insulin. The role of negative cooperativity *in vivo* is uncertain. The most obvious functon would be to maintain insulin sensitivity at low hormone concentrations, yet to buffer against higher concentrations of insulin such as might occur following a burst of secretory activity by the B-cell. In most instances, the magnitude of the response of a target tissue to insulin is propor-tional to the number of receptor sites occupied by insulin. However, a maximal response is normally achieved when only a minority (2-10 per cent) of the receptor sites are occupied. This finding suggests the presence of 'spare receptors' and indicates that the number of receptors on the cell surface that are occupied by insulin is not a rate-limiting step in insulin action. The biological significance of 'spare receptors' is not certain.

The major features of the insulin receptor have been compared in a variety of species which span approximately 400 million years of evolution. Several important properties of the receptor, including pH and temperature dependence and negative cooperativity, were found to be virtually identical in all of the species examined. In addition, the specificity of the receptor appears to be highly conserved since the receptors from birds, mammals, amphibians and bony fish have similar absolute and relative affinities for insulins of different species. These findings suggest that the insulin receptor has been well conserved throughout evolution. Thus, since insulin and its receptor have changed relatively little during evolution, it is possible that the insulin-receptor complex is more than a simple recognition system and that it may, in addition, fulfil some specific catalytic function in the plasma membrane.

The localisation of the insulin receptor as an integral part of the plasma membrane has hampered its biochemical characterisation, since it has been difficult to solubilise and separate from other membrane components. Various detergent treatments have, however, given a soluble receptor preparation which retains insulin-binding properties. This solubilised receptor is a large glycoprotein, with a molecular weight in the region of 350,000 daltons. It is composed of two major subunits, the α-subunit (130,000 daltons) and the β-subunit (95,000 daltons) and these may be linked in pairs by disulphide bonds to form an immunoglobulin-like structure of the type $(\alpha\beta)_2$. It is not known whether the insulin receptor has any catalytic capability, although the demonstration that the interaction of insulin with the receptor leads to the phosphorylation of the β-subunit suggests that this is likely.

Insulin receptors are normally free to diffuse laterally within the plane of the plasma membrane. However, when they bind insulin there is an increase in the affinity of the free mobile receptors for each other and they may aggregate into immobile patches (patching). Areas of membrane rich in aggregates of receptor bound insulin may then become internalised by the process of endocytosis. The internalised areas of membrane are subjected to lysosomal attack and the insulin is rapidly degraded. The receptor, however, appears more resistant to degradation and it may be cycled back to the plasma membrane. This process (receptor-mediated endocytosis) plays an important role in regulating the number of receptors on the surface of the cell and it represents an important pathway of insulin degradation.

Although insulin receptors were thought to be localised largely on the plasma membrane of target cells, they have now been demonstrated on intracellular membranes such as those of the nucleus, endoplasmic reticulum and Golgi apparatus. The binding of insulin to the nuclear membrane is rapid, reversible, of high affinity and hormone specific, characteristics which are similar to those of the plasma membrane receptor. The nuclear membrane appears to have a similar total insulin binding capacity to that of the plasma membrane, and the binding capacity of both appears to be regulated by the extracellular insulin concentration. Several studies have indicated that insulin can enter target cells and interact directly with these internal receptors. These observations raise the possibility that insulin may regulate certain intracellular events such as protein, DNA and RNA synthesis by entering the cell and interacting with intracellular receptors. This would account for the delayed appearance of these effects following insulin exposure since the uptake of insulin into cells is a relatively slow process. However, the physiological significance of intracellular insulin receptors is unknown at the present time.

Regulation of the number of insulin receptors on the plasma membrane of a target cell and alterations in the affinity of such receptors for insulin may provide a mechanism whereby the cell can alter its responsiveness to the hormone. Indeed it appears likely that, rather than being passive recipients of hormonal stimuli, cells act continuously to regulate their responsiveness to stimulation. One of the most important regulators of insulin receptor concentration is insulin itself and prolonged elevations in the concentration of insulin in the circulation may lead to a concentration related reduction in the number of receptors on the surface of all target cells. This 'down-regulation' of insulin receptors may be physiologically important in preventing over stimulation of target cells by insulin. However, 'down-regulation' of receptors, if persistent, would produce a right-ward shift in the insulin dose-response curve, i.e. it would lead to insulin resistance (Figure 4.1).

Insulin Resistance

Insulin resistance is said to occur when normal concentrations of insulin produce a less than normal biological response and there are a number of important clinical disorders and metabolic states which share insulin resistance as a common feature (Table 4.2). Generally these insulin resistant states are characterised by an increase in the basal levels of circulating insulin (hyperinsulinism) and a reduced effectiveness of exogenous or endogenous insulin. The resistance may be severe enough to produce glucose intolerance although this is not always seen. Since insulin travels from the B-cell to a target cell through the circulation, events at any one of these loci could influence the ultimate action of the hormone. Table 4.3 lists some of the known causes of insulin resistance in man. Of these, target tissue defects appear to be the most important in producing clinically important insulin resistance states.

Table 4.2: Insulin Resistant States Characterised by Changes in Receptor Number and/or Affinity

State	Receptor number	Receptor affinity
Obesity	Decrease	No change
Cushing's syndrome (glucocorticoid excess)	No change	Decrease
Acromegaly (growth hormone excess)	Decrease	Increase
Type 2 diabetes	Decrease	No change
Insulinoma	Decrease	No change or increase

Table 4.3: Possible Causes of Insulin Resistance

1. Circulating insulin antagonists
   (a) Elevated levels of counter regulatory hormones growth hormone, placental lactogen, cortisol, glucagon or catecholamines
   (b) Anti-insulin antibodies
   (c) Anti-insulin receptor antibodies

2. Target tissue defects
   (a) Insulin receptor defects decreased affinity, decreased number
   (b) Post-receptor defects defective coupling, defective response mechanism

Circulating Insulin Antagonists

Many of the actions of insulin on target tissues are opposed by other hormones and there are several clinical syndromes in which elevated levels of these hormones can induce an insulin resistant diabetic state. In many of these syndromes the high circulating level of hormone produces changes in the insulin receptor which are responsible for the resistance. Thus glucose intolerance is frequently observed in patients with Cushing's syndrome or in patients given glucocorticoids for treatment of other diseases. The major factor in this form of insulin resistance is a decrease in the affinity of the insulin receptor for insulin caused by the high plasma glucocorticoid levels (Table 4.3). In addition, acromegaly is commonly associated with hyperinsulinaemia and insulin resistance. This appears to be related to a growth hormone induced decrease in receptor concentration and an increase in the affinity of the remaining receptors for insulin. These changes result in normal amounts of insulin being bound at basal insulin concentrations but a decrease in insulin binding at high insulin concentrations.

Anti-insulin antibodies develop in most patients treated over long periods with insulin but it is only rarely that these antibodies cause clinically significant insulin resistance. In recent years, however, a syndrome has been described in which these patients develop antibodies of the IgG class against the insulin receptor. The antibodies bind to the insulin receptor preventing insulin's interaction and producing severe insulin resistance and diabetes. These antibodies have been very useful as probes of the structure and function of the insulin receptor. Of particular significance in this context was demonstration that the interaction of the antibody with the receptor mimics all of the actions of insulin except those related to the promotion of cell growth. This observation provides strong evidence that all the effects of insulin, other than those related to the promotion of growth, are the result of insulin's interaction with a single class of receptors.

Target Tissue Defects

Insulin exerts its biological effects by binding to its specific cell surface receptor. The interaction of insulin with its receptor leads to the generation of one or more signals that interact with the effector units that mediate the many biological responses. Insulin's action on a target cell thus involves a cascade of events, and abnormalities anywhere along this sequence could lead to insulin resistance. Insulin resistance due to target tissue defects can be considered in terms of a decreased sensitivity and/or a decreased responsiveness of the tissue to insulin (Figure 4.1).

The former is characterised by a shift to the right in the insulin dose-response curve with no change in maximal activity and may result from a decrease in receptor number or in the affinity of the receptor for insulin. The latter is characterised by a reduced maximal activity and may be caused by a post-receptor defect.

A decreased number of cell surface insulin receptors, associated with insulin resistance, has been described in several clinical conditions including obesity and type 2 diabetes, although the insulin resistance in type 1 diabetes does not appear to involve such a decrease. The resistance to insulin in obesity which is the commonest form of insulin resistance in man is thought to be related to the marked reduction (30-60 per cent) in the number of insulin receptors on target cells such as adipocytes. The reduction in receptor number appears to be a result of the down regulation of the receptor number in response to the hyperinsulinism which is a consequence, at least initially, of the excessive food intake. Effective treatment of the obese subject by caloric restriction, resulting in significant weight loss, is associated with a fall in the

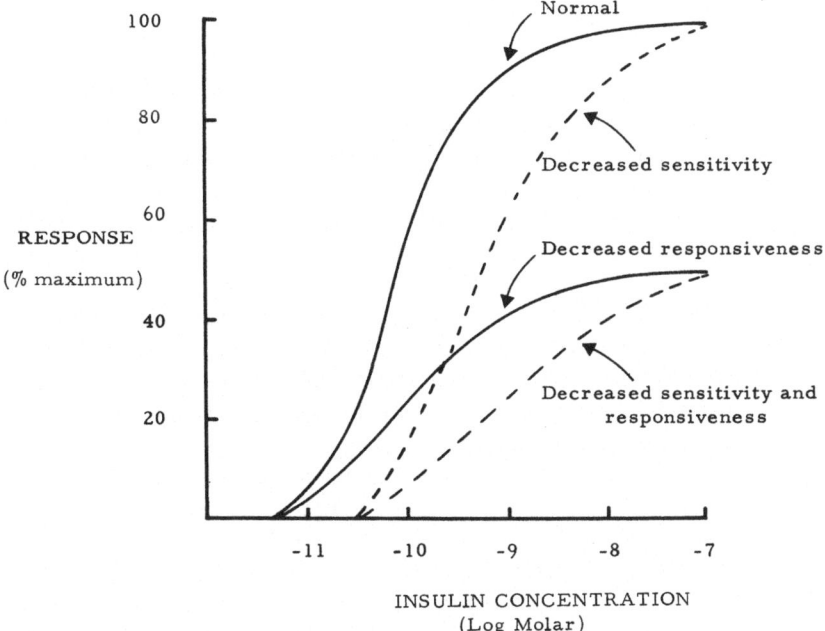

Figure 4.1: Possible Tissue Responses in Insulin Resistance

serum insulin concentration and a return of the insulin receptor number to normal.

The loss of insulin receptors appears to account for the loss of insulin sensitivity in obesity and type 2 diabetes when the insulin resistance is relatively mild. However, as insulin resistance becomes more severe and there is a loss of insulin responsiveness, then post-receptor defects become more pronounced, and in those patients with severe insulin resistance, the post-receptor defects predominate. The nature of the post-receptor defects in diabetes and obesity have not yet been determined. A post-receptor defect involving the coupling between insulin receptor complexes and the glucose transport system has however been demonstrated in fibroblasts from two patients with diabetes associated with Leprechaunism. The insulin resistance in these patients appeared to be due entirely to the post-receptor defects as the patient's circulating plasma insulin was chemically and biologically normal, there were no circulating insulin antagonists in the plasma and insulin receptor concentration appeared normal.

Insulin Action

Insulin is an absolute requirement for normal growth and development in man and is essential for the maintenance of life since its absence leads to rapid tissue wasting and death. It is a potent hormone that influences the metabolism and function of most tissues although the key target tissues appear to be liver, muscle and adipose tissue. Many of the physiological effects of insulin in man are exerted at a plasma concentration in the region of 10^{-10} M and it has major effects on carbohydrate, lipid and protein metabolism. The major function of insulin in man is to coordinate the relative rates of fuel uptake and storage and fuel mobilisation with his varying needs and with the availability or lack of availability of fuel from the diet. In particular, insulin promotes the rapid uptake and utilisation or storage of fuels following a meal and permits their controlled release when they are subsequently required. In this way, the fuel molecules in the blood may be maintained at a concentration appropriate to the competing needs of all tissues and major changes in this concentration can normally be avoided.

Insulin is unique among the hormones produced by man in the wide range of tissues, and processes within those tissues, that it affects (Table 4.4). For convenience these effects can be divided into those which occur rapidly (sec-min) and those which take a relatively long time

Table 4.4: Effects of Insulin on Target Tissues

Effect	Tissue
(a) Rapid effects	
Increased membrane transport of glucose	Muscle, Adipose
Increased membrane transport of amino acids	Muscle, Adipose, Liver
(b) Intermediate effects	
(i) Carbohydrate metabolism	
Increased glycogen synthesis	Muscle, Liver
Decreased glycogenolysis	Muscle, Liver
Increased glycolysis	Muscle, Liver, Adipose
Decreased gluconeogenesis	Liver
(ii) Lipid metabolism	
Increased lipogenesis	Liver, Adipose
Increased esterification	Liver, Adipose
Decreased lipolysis	Adipose
Increased cholesterol synthesis	Liver
Decreased ketogenesis	Liver
Increased utilisation of dietary lipid	Liver, Adipose
Decreased fatty acid oxidation	Liver, Adipose
(iii) Protein metabolism	
Increased protein synthesis	Liver, Muscle, Adipose
Decreased proteolysis	Liver, Muscle
(c) Long-term effects	
Promotion of cell growth	
Promotion of cell division	
Promotion of DNA synthesis	
Promotion of RNA synthesis	

(min-hours) to manifest themselves. In addition, there are those effects which occur on a time scale which is between these two extremes. The rapid and intermediate effects of insulin are generally dependent on the continued presence of insulin and their magnitude is proportional to the insulin concentration. The long-term effects may, however, appear after insulin levels have fallen back to the pre-stimulatory value. There also appear to be differences in the sensitivities of the effects to insulin since the long-term effects appear to require much higher concentrations of insulin than the short-term effects.

Rapid Effects of Insulin

One of the first detectable actions of insulin on many cells is the rapid stimulation of the plasma membrane transport of a variety of charged and neutral substances. Thus, insulin promotes the transport of cations

(Na^+, K^+, and Ca^{2+}), anions (PO_4^{3-}), certain amino acids and D-glucose into the major insulin-sensitive tissues such as muscle and adipose tissue, although it is without effect on the transport of glucose into liver. Of all the known insulin-sensitive transport systems, the D-glucose transport system has been the most extensively studied.

Hexose transport in muscle and adipose tissue is stereospecific and operates by facilitated diffusion both in the basal and insulin-stimulated state. Thus, the concentration gradient of D-glucose across the plasma membrane provides the driving force for net influx. The effect of insulin on transport occurs rapidly (seconds-minutes) following the addition of insulin and deactivation following the removal of insulin is also rapid although it does occur more slowly than activation. Kinetic studies have shown that insulin stimulates glucose transport by increasing the apparent maximum activity (V_{max}) of the glucose carrier with little or no change in the apparent affinity (K_m) of the carrier for glucose. A change in V_{max} of the system could be due to an effect of insulin on the activity of a fixed number of transporter units or on the number of these units. Recently, methods have become available which allow an assessment to be made of the number of glucose transport units in subcellular fractions of cells. These methods depend on the observation that cytochalasin B can compete with D-glucose for binding to glucose transport units. Using these methods, it has been shown that insulin stimulates glucose transport in muscle and adipose tissue by promoting the translocation of functional glucose transport units from a specific intracellular membrane pool, possibly the Golgi complex, to the plasma membrane. The extent of translocation appears to correlate with the increase in glucose uptake rate in response to insulin. Moreover, the effect is rapid and reversible and the time scales of the changes are similar to those observed for the insulin-induced alterations in glucose uptake rate. Thus, while it seems likely that insulin promotes the uptake of glucose in muscle and adipose tissue by increasing the number of transport units in the plasma membrane, it is not known to what extent these observations can be related to the other insulin-sensitive transport processes.

Intermediate Effects of Insulin

Insulin has many effects on the metabolism of carbohydrates, lipids and amino acids which usually occur within a few minutes of hormone exposure and are produced by insulin concentrations in the physiological range. In many instances they involve changes in the activities of existing enzyme molecules brought about by reversible covalent modifications of the enzyme structure. This usually involves the phosphorylation/dephosphorylation of specific serine or threonine residues in the enzyme, the phosphorylation being catalysed by a protein kinase and the dephosphorylation being catalysed by a phosphoprotein phosphatase. In some instances the phosphorylated enzyme is active while in others the dephosphorylated enzyme is active (Table 4.5).

Table 4.5: Effects of Insulin on Enzyme Activity

Enzyme	Effect of insulin on activity	Molecular basis of activity change
(a) Carbohydrate metabolism		
Glycogen synthase	Increases	Dephosphorylation
Phosphorylase	Decreases	Dephosphorylation
Phosphorylase kinase	Decreases	Dephosphorylation
Pyruvate dehydrogenase	Increases	Dephosphorylation
Pyruvate kinase	Increases	Dephosphorylation
Fructose-6-phosphate-2-kinase	Increases	Dephosphorylation
Phosphoprotein phosphatase inhibitor 1	Decreases	Dephosphorylation
(b) Lipid metabolism		
Triacylglycerol lipase	Decreases	Dephosphorylation
Hydroxymethyl glutaryl-CoA reductase	Increases	Dephosphorylation
Hydroxymethyl glutaryl-CoA reductase kinase	Decreases	Dephosphorylation
Acetyl-CoA carboxylase	Increases	Phosphorylation
ATP-citrate lyase	?	Phosphorylation
Glycerol phosphate acyltransferase	Increases	Dephosphorylation
Diacylglycerol acyltransferase	Increases	Dephosphorylation
(c) Phosphoprotein phosphatase and protein kinase enzymes		
Glycogen synthase phosphatase	Increases	Peptide factor
Pyruvate dehydrogenase phosphatase	Increases	Peptide factor
Cyclic AMP-independent protein kinase	Increases	?
Cyclic AMP-dependent protein kinase	Decreases	Peptide factor
(d) Membrane enzymes		
Na/K-ATPase	Increases ?	Phosphorylation
Low K_m cyclic AMP phosphodiesterase	Increases	Phosphorylation
Ca-ATPase	Decreases	Dephosphorylation ?

Carbohydrate Metabolism. The action of insulin on carbohydrate metabolism normally ensures that glucose is rapidly removed from the blood and utilised by tissues when the blood glucose concentration is raised above the fasting level. This important function serves to keep to a minimum the period of hyperglycaemia which follows the dietary intake of a carbohydrate meal. To achieve this, insulin promotes the uptake of glucose by muscle and adipose tissue although not by the liver. It also promotes the storage and/or utilisation of glucose by all three of these tissues. The effect of insulin on these metabolic processes is rapid and largely involves the activation or inactivation of the enzymes that control the rate of flux of glucose through the relevant metabolic pathways.

The storage of glucose by muscle and liver is achieved by the formation of glycogen, and insulin increases glycogen synthase activity, the rate determining step of the glycogen synthesis pathway (glycogenesis). In addition, it also inhibits glycogen breakdown (glycogenolysis) by decreasing phosphorylase activity, the rate-limiting step of glycogenolysis. When the glycogen stores are full, any excess glucose is either oxidised to CO_2 via glycolysis and the Krebs cycle and thereby used to drive ATP synthesis, or else it is converted via glycolysis and lipogenesis into fatty acids which are subsequently esterified and stored in adipose tissues as triglyceride. In general, adipose tissue and liver are major sites of lipogenesis from glucose while muscle tends to oxidise excess glucose to CO_2. In order to promote the utilisation of glucose via glycolysis insulin increases the activity of the rate-limiting enzymes, pyruvate kinase, pyruvate dehydrogenase and possible 6-phosphofructo-1-kinase thereby ensuring the irreversible conversion of glucose into acetyl-CoA. The subsequent conversion of acetyl CoA into fatty acids (lipogenesis) is promoted via the activation by insulin of acetyl-CoA carboxylase and possibly citrate lyase. Insulin also promotes the esterification of the newly formed fatty acids by increasing the availability of glycerol phosphate, the precursor for the glycerol backbone of the triglycerides. In addition, it may stimulate diacylglycerol acyltransferase and glycerol phosphate acyltransferase activities.

Lipid Metabolism. Insulin plays an essential role in regulating the blood and tissue content of several classes of lipid and it ensures that excess nutrients, be they glucose, fatty acids or amino acids, are converted into triglyceride and stored in adipose tissue. Insulin regulates the blood content of lipid by ensuring (a) that dietary triglyceride is rapidly removed from the circulation, (b) that fatty acids are not released from stored triglyceride until needed, (c) that ketone body production by the liver is minimal. In diabetes, when insulin levels are low or absent these actions of insulin disappear and blood lipid levels increase, there being particularly dramatic increases in fatty acid and ketone body levels.

The disposal of dietary triglyceride depends largely on the activity of lipoprotein lipase in a number of tissues. This enzyme is an extracellular enzyme released from tissues such as adipose tissue and it hydrolyses chylomicron and very low density lipoprotein (VLDL) triglyceride, thereby making it available to tissues. Insulin promotes the synthesis and release of the enzyme from adipose tissue and thereby increases the rate of dietary triglyceride uptake.

The inhibitory effect of insulin on the release of fatty acids from adipose tissue triglyceride is related partly to its ability to inhibit triglyceride lipase, the rate-limiting enzyme of triglyceride breakdown, and also to its ability to stimulate the esterification of fatty acids. This is due in part to the increased uptake and utilisation of glucose by adipose tissue which occurs under the influence of insulin and the consequent increased production of glycerol phosphate, a necessary component of the esterification process.

Insulin's action in inhibiting ketone body production (ketogenesis) depends in part on the maintenance of low circulating levels of fatty acids since elevated levels are a prerequisite for ketogenesis. In addition, insulin activates acetyl-CoA carboxylase in the liver and thereby ensures high intracellular levels of malonyl-CoA which prevents the entry of fatty acids into the mitochondria, the site of ketone body production.

Protein Metabolism. Nitrogen imbalance, reflected as tissue wasting, increased nitrogen excretion and increased concentrations of blood amino acids, is often one of the striking metabolic disturbances associated with type 1 diabetes. These changes are the result of an imbalance in protein turnover, with the rate of protein degradation exceeding the rate of synthesis. Insulin administration corrects the nitrogen imbalance in diabetes and this is related to its ability to affect both the synthesis and degradation of protein in many tissues including heart, skeletal muscle and liver. The turnover of skeletal muscle and liver protein is particularly important in diabetes because it provides the bulk of the amino acid substrates for gluconeogenesis thereby contributing to the hyperglycaemia of the condition.

The intracellular degradation of proteins is an essential part of the system responsible for regulating the cellular content of both exogenous and endogenous proteins. In addition, it functions to remove abnormal proteins that may be formed by mutation, ageing or chemical modification. The amino acids released are available for new protein synthesis, energy production and the synthesis of essential nitrogen-containing compounds. The major portion of the protein degrading activity of the cell occurs in the lysosomes, subcellular organelles which are rich in proteases. Insulin inhibits protein degradation in many tissues including skeletal and heart muscle and liver with a time scale which varies from minutes to hours depending on the tissue. This effect may be related to an inhibition by insulin of autophagic lysosomal vacuole formation and the inactivation of lysosomal protease, although the molecular basis of insulin's action on proteolysis is not known in detail.

Insulin has a general anabolic effect on protein metabolism in muscle, liver and adipose tissue stimulating the synthesis of a wide range of proteins. In addition, in liver, insulin increases the amount of a number of specific proteins including the glycolytic enzymes glucokinase, 6-phosphofructo-1-kinase and pyruvate kinase. These effects of insulin on protein synthesis appear over a time scale of five minutes to several hours depending on the tissue and the protein being synthesised and may involve several independent mechanisms. Thus insulin increases the membrane transport of several (but not all) amino acids and it increases the ability of ribosomes to translate messenger RNA (translational effect). In addition, it stimulates specific mRNA synthesis (transcriptional effect) and thereby also promotes the synthesis of specific proteins. Whilst the molecular basis of insulin's action on protein synthesis has not been defined in detail it has been demonstrated that insulin increases the phosphorylation of ribosomal protein factor 6 of the 60S ribosomal subunit. The phosphorylation/dephosphorylation of this protein may conceivably play some role in regulating the interaction of ribosomes with mRNA.

Long-term effects of insulin

Insulin promotes the growth, differentiation and proliferation of many cells in culture including those from skin, liver, adipose tissue and mammary gland. In general, these effects of insulin involve the stimulation of RNA and/or DNA synthesis and often require relatively high concentrations of insulin (10^{-8}-10^{-6}M). They usually become manifest several hours (4-24 hours) after the initial exposure to insulin and insulin must be present for an extended time period. In some cell types insulin alone will stimulate DNA synthesis and/or growth, while in others it acts synergistically with growth factors to promote cell proliferation. These differences in growth response to insulin probably depend to a large extent on the particular processes that are growth limiting in different cell types under different conditions.

It has been suggested that the stimulatory effect of insulin on growth may arise as a consequence of the acute anabolic effects of insulin on cellular metabolism. Thus, insulin promotes the uptake and utilisation or storage of essential nutrients such as amino acids and glucose by cells and this may lead to cell growth. A major problem with this explanation is that the effects of insulin on growth are usually only observed at high concentrations of insulin while its acute metabolic effects on the same cells are often optimal at low concentrations. This has raised the possibility that insulin might stimulate growth by interacting with receptors for one of the other growth-promoting hormones. Indeed a family of peptide hormones, with a spectrum of growth-promoting activities similar to those of insulin, has been isolated from human plasma. The proteins in this family have a molecular weight in the range 7,000-8,000 and include the insulin-like growth factors (IGFs I and II) and the somatomedins (A and C). Although they are immunologically distinct from insulin, they are capable of eliciting all of insulin's biological actions. In general, they are much more potent (10-1000 times) than insulin in their growth-promoting actions and considerably less potent (100 times) than insulin in their acute metabolic effects. These peptides appear to have their own high affinity receptors on target cells and the receptors are distinct from the insulin receptor. They do, however, interact with low affinity with the insulin receptor and insulin has been shown to interact with the growth factor receptors. At least part of this cross-reactivity can be accounted for by structural similarities between the insulin molecule and the growth factor molecule since IGFs I and II share about 50 per cent sequence homology with insulin. These observations suggest that the growth-promoting actions of insulin may be mediated in part by the interaction of insulin with the receptors for the insulin-like growth factors rather than with insulin receptor *per se*. Evidence in favour of this hypothesis has been provided by the observation that anti-insulin receptor antibodies do not affect the growth response to insulin but they do affect the acute metabolic effects. It appears likely therefore that some of the effects of insulin on cellular growth are attributable to interaction of insulin with IGF/somatomedin receptors while other effects on growth may be caused by binding of insulin to its own high-affinity receptor.

It is not known how the interaction of insulin with these receptors leads to the growth response but it has been suggested that the delay in the appearance of the effect may be related to the need for insulin to enter the cell and reach the nucleus before it can act. The presence of insulin receptors on the nuclear membrane and the observation that insulin can enter target cells by the process of receptor mediated endocytosis both support this suggestion. However, it has yet to be established whether the growth-promoting actions of insulin are related to the cellular uptake of the hormone.

The extent to which the growth effects of insulin, observed in cell culture, occur *in vivo* during development is uncertain, although there are a number of clinical examples which suggest that insulin is an important growth-regulation hormone during human foetal development. Thus, infants born to diabetic mothers with hyperglycaemia late in pregnancy often have large and excessive body weight. This has been attributed to hyperinsulinaemia in the foetus caused by the maternal hyperglycaemia. Conversely, infants with insulin deficiency at birth are characterised by small body size for gestational age. These growth abnormalities can be considered in terms of direct effects of insulin on growth although it may be that some of the growth-promoting actions of insulin *in vivo* are indirect.

Molecular Basis of Insulin Action

The mechanism whereby the interaction of insulin with its plasma membrane receptor leads to the appropriate cellular response is not yet understood. However, it is likely that the binding of insulin to the receptor triggers a conformational change in the receptor which results in the generation of a transmembrane signal. The signal molecule is then responsible for mediating the intracellular actions of insulin.

One of the major problems which has prevented an understanding of the mechanism of insulin action has been the difficulty, until recently, of obtaining effects of insulin on broken cell preparations. In addition, since insulin exerts a large number of diverse effects on cells with different temporal sequences (Table 4.4) and different sensitivities to insulin, it is difficult to fit them all into any comprehensive theory of insulin action. It may be that the diverse effects are mediated by different signal molecules and that insulin or a fragment of the insulin molecule could act as one of its own intracellular signals. Indeed, if one considers the variety of actions of insulin it would appear likely that insulin should have several signal molecules. Thus, for many of its effects on intermediary metabolism insulin's action is generally antagonistic to that of other hormones and one might anticipate that its action would involve the same signal molecules that are used by other hormones. Indeed, studies have implicated cyclic AMP, cyclic GMP and calcium as second messengers of insulin action. However, the rapid effects on membrane transport and the long-term effects on DNA and RNA biosynthesis are independent of those of other hormones and are likely therefore to involve novel signal molecules.

Insulin and Cyclic Nucleotides

Many of the effects of insulin are antagonistic to those hormones which exert their effects by increasing intracellular cyclic AMP concentrations. It appeared likely therefore that insulin might exert its effects by lowering intracellular cyclic AMP concentrations. Indeed, it has been shown that insulin will reduce the cyclic AMP concentration in liver, muscle and adipose tissue. These effects may be related to the stimulation of a low K_m cyclic nucleotide phosphodiesterase by insulin, although inhibitory effects of insulin on adenylate cyclase activity have also been observed. The change in intracellular cyclic AMP concentration induced by insulin in adipose tissue appears to be physiologically significant since it is paralleled by changes in cyclic AMP-dependent protein kinase activity in this tissue. In addition to affecting cyclic AMP metabolism, insulin has also been shown to increase cyclic GMP levels in some tissues, possibly by activating guanylate cyclase, although the physiological significance of this change is not known.

The Role of Calcium in Insulin Action

It has been suggested that calcium may be the signal molecule responsible for mediating the intracellular actions of insulin, and that the interaction of insulin with its receptor triggers an increase in intracellular calcium which is responsible for the subsequent changes in metabolism. This suggestion is attractive, since some of the intracellular events that are affected by insulin are also affected by calcium and many agents that have insulin-like effects on tissue are known to increase cytoplasmic calcium concentrations.

Cells normally maintain a large electrochemical gradient for ionised calcium between the cytoplasm ($<10^{-6}$ M) and the extracellular environment ($>10^{-3}$ M). Alterations in the plasma membrane components responsible for maintaining this gradient could therefore have marked effects on the cytoplasmic calcium concentration. One such component is the active extrusion of calcium from the cell by a high affinity calcium-stimulated ATPase (Ca^{2+}-ATPase). The activity of this enzyme in adipocytes has been shown to be inhibited by physiological concentrations of insulin suggesting that direct regulation of intracellular calcium homeostasis may indeed represent an important event in the mechanism of action of insulin. However, since some of the actions of insulin do not appear to involve calcium sensitive enzymes and others are opposite to those induced by calcium, this cannot account for all the actions of insulin. The demonstration that insulin inhibits the phosphorylation and hence the activity of the Ca^{2+}-ATPase suggests that the effect of insulin on intracellular calcium homeostasis may be secondary to alterations in the activity of a protein kinase or a phosphoprotein phosphatase. Thus while calcium may be important in mediating some of the effects of insulin it does not appear to be the 'second messenger' of insulin action.

Role of Insulin Uptake in Insulin Action

The presence of insulin receptors on intracellular membranes such as those of the nucleus and endoplasmic reticulum, and the demonstration of the uptake of insulin into the cell, has led to the suggestion that insulin or a part of the molecule may exert its effects intracellularly. This suggestion would account for the difficulty in demonstrating an intracellular messenger of insulin action since insulin itself or a product of its breakdown might be the second messenger. Indeed it appears that synthetic fragments of the insulin B-chain have some biological activity but do not interact with the insulin receptor. The weight of evidence however argues against this suggestion. Thus many of the effects of insulin occur within a time scale that is too rapid for internalisation to occur. In addition, many of the cellular effects of insulin are produced by antibodies against the insulin receptor and it is difficult to see how these could be degraded to give rise to the appropriate intracellular messenger. Furthermore, inhibitors of the uptake and degradation of insulin do not appear to block the actions of the hormone. Finally the demonstration of a peptide mediator of insulin action makes it no longer necessary to suggest that insulin must act as its own second messenger.

Peptide Mediator of Insulin Action

Although insulin has effects on the classical second messengers of hormone action, such as cyclic nucleotides and calcium, it has always been a possibility that insulin might have its own distinct second messenger. Recently evidence has been obtained in support of this possibility. Thus, it has been shown that exposure of skeletal muscle, liver and adipocytes to physiological concentrations of insulin results in the production of a low molecular weight 'messenger' that mimics the action of insulin on a number of enzymes. To date it has been shown to affect the following insulin responsive enzymes; pyruvate dehydrogenase, cyclic AMP phosphodiesterase, glycogen synthase, Ca^{2+}-Mg^{2+}-ATPase, cyclic AMP-dependent protein kinase and certain phosphoprotein phosphatases.

The molecular weight of the 'messenger' is in the region of 1,000-

2,000 daltons. It is heat and acid stable and appears to be a polypeptide since it is rapidly destroyed by proteases. It is produced rapidly, following the interaction of insulin with its plasma membrane receptor and it may be generated from a larger membrane protein by proteolytic cleavage. In order to account for the rapid generation of the messenger, it has been suggested that the interaction of insulin with its plasma membrane receptor leads to the activation of a specific plasma membrane protease. The protease then acts on its substrate releasing the polypeptide messenger.

Thus, a molecule has been isolated from all of the major insulin-responsive tissues which has the features of other second messengers of hormone action in that it is small, heat stable, rapidly produced in the plasma membrane and readily destroyed. It seems likely, therefore, to be a second messenger of insulin action. The messenger is able to stimulate the dephosphorylation and activation of certain insulin-sensitive systems such as pyruvate dehydrogenase and glycogen synthase by activating specific phosphoprotein phosphatases or by inactivating cyclic AMP-dependent protein kinases. These properties are sufficient to account for those actions of insulin which involve the dephosphorylation of key proteins. The occurrence of the messenger does not, however, appear to provide an explanation of those effects of insulin which involve increased phosphorylation of key proteins. It may be that an additional messenger is responsible for these actions. Indeed, some evidence has been obtained for the production of a phospho protein phosphatase inhibitory factor in insulin-treated muscle tissue. In addition, since the messenger appears to arise from a membrane protein by proteolysis it is difficult, at this stage, to envisage how enough messenger could be generated. Thus, unlike other second messenger systems, there does not appear to be any amplification at the messenger generation step. This objection could, however, be overcome by the demonstration that the messenger itself possesses some form of catalytic activity.

Unifying Hypothesis of Insulin Action

The number of theories concerning the molecular basis of insulin's action is almost as great as the number of different actions of the hormones on target tissues. Table 4.6 lists the molecular events which have been proposed to account for insulin's action. They are all components of the highly integrated regulatory systems that exist within animal cells to enable them to respond to stimuli such as hormones and neurotransmitters in their environment (Figure 4.2). Since many of the

Table 4.6: Possible Primary Sites of Insulin Action

Molecular Event	Mechanism of Insulin's Action
1. Lowering of intracellular cyclic AMP concentrations and the consequent reduction of cyclic AMP-dependent protein kinase activity	(a) Inhibition of adenylate cyclase (b) Activation of cyclic nucleotide phosphodiesterase
2. Raising of intracellular cyclic GMP concentrations	Activation of guanylate cyclase
3. Increasing intracellular and/or mitochondrial calcium concentrations	Increasing activity of membrane calcium translocases
4. Activation of phosphoprotein phosphatase activity	?
5. Activation of specific cyclic nucleotide independent protein kinases	?

actions of insulin are antagonistic to those of other hormones it is hardly surprising that the components of these regulatory systems show changes in response to insulin which are, in general, opposite to those induced by other hormones. Indeed, since these are highly integrated regulatory systems, one would be surprised not to see many of these changes. The question thus becomes which, if any, of these events are primary, or is there some other primary event(s) which triggers the observed changes?

Many of the actions of insulin are antagonistic to those of adrenaline and/or glucagon and involve reciprocal changes in the activities of key enzymes. These enzymes are all regulated by reversible phosphorylation at specific sites in the molecule. In some instances there may be phosphorylation at additional sites (multisite phosphorylation) which may not lead directly to a change in activity but in general, the degree of phosphorylation determines the activity. Phosphorylation may lead to the activation or inactivation of an enzyme. The degree of phosphorylation is controlled by the relative activities of the phosphorylating enzyme (a protein kinase) and the dephosphorylating enzyme (a phosphoprotein phosphatase). In general the effects of glucagon and adrenaline are mediated by an increased protein kinase activity (calcium and/or cyclic AMP dependent) and a reduced phosphoprotein phosphatase activity, i.e. an increased phosphorylation of the key enzyme. The reciprocal changes in protein kinase and phosphoprotein phosphatase activities are necessary to ensure that a futile cycle is not set up.

Not surprisingly the antagonistic effects of insulin on these key

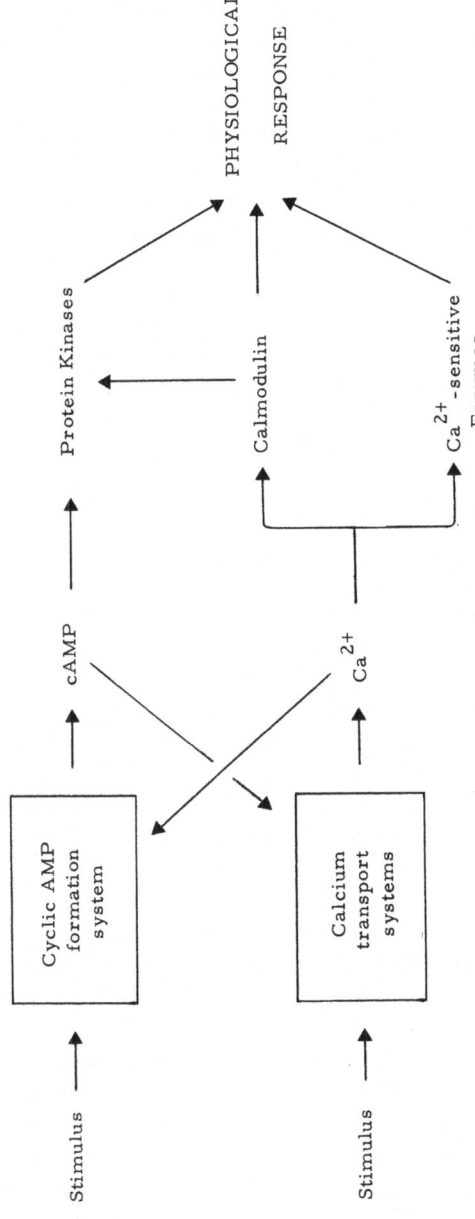

Figure 4.2: Regulatory Systems used by Animal Cells to Enable them to Respond to Extracellular Stimuli

enzymes are mediated by dephosphorylation and insulin has been shown to inhibit cyclic AMP-dependent protein kinase activity and to stimulate certain phosphoprotein phosphatase enzymes (Table 4.7).

Table 4.7: Effects of Insulin on Protein Phosphorylation/Dephosphorylation Enzymes

Enzyme	Effect of insulin on activity
(a) Protein kinases	
Phosphorylase kinase	decreases
Hydroxymethylglutaryl-CoA reductase kinase	decreases
Cyclic AMP-dependent protein kinase	decreases
Cyclic AMP-independent protein kinase	increases
Pyruvate kinase	increases
Fructose-6-phosphate-2 kinase	increases
(b) Phosphoprotein phosphatases	
Glycogen synthase phosphatase	increases
Pyruvate dehydrogenase phosphatase	increases

Other actions of insulin also appear to involve changes in the phosphorylation state of key enzymes and other functional proteins (Table 4.8). In these cases however, insulin appears to increase the activity of specific cyclic AMP-independent protein kinases and promote the phosphorylation of the substrate. In general, these actions are not opposed by glucagon or adrenaline and where they are (acetyl-CoA carboxylase) then insulin appears to phosphorylate different sites in the molecule.

Thus it is possible to account for many, if not all, of the effects of insulin on intermediary metabolism on the basis of insulin-induced changes in the activities of certain protein kinases and phosphoprotein phosphatases.

It is not yet known how the interaction of insulin with its plasma membrane receptor leads to the changes. However, the isolation of a peptide which appears to mediate the effects of insulin on protein kinase and phosphoprotein phosphatase activities and the demonstration of plasma membrane protein kinase which is sensitive to insulin suggest that we may be near to solving this problem. The recent demonstration that insulin promotes the phosphorylation of the β-subunit of its own receptor may be of significance in this context.

Table 4.8: Effects of Insulin on Protein Phosphorylation

(a) Proteins whose phosphorylation is decreased by insulin

Glycogen synthase
Phosphorylase
Phosphorylase kinase
Pyruvate dehydrogenase
Pyruvate kinase
Fructose-6-phosphate 2-kinase
Phosphoprotein phosphatase inhibitor 1
Glycerolphosphate acyltransferase
Diacylglycerol acyltransferase
Triacylglycerol lipase
Hydroxymethylglutaryl-CoA reductase
Hydroxymethylglutaryl-CoA reductase kinase
Ca^{2+}-ATPase
Liver plasma membrane integral proteins
    (mol.wt. 140,000, 80,000)

(b) Proteins whose phosphorylation is increased by insulin

Acetyl-CoA carboxylase
ATP-citrate lyase
Cyclic AMP phosphodiesterase (low K_m enzyme)
Na/K-ATPase (sarcolemma protein mol.wt. 16,000)
Liver plasma membrane peripheral proteins
    (mol.wt. 28,000, 14,000)
Ribosomal protein S6 - 40S subunit
Insulin receptor (β subunit)

Insulin Metabolism

Following its release from the pancreas or its intravenous administration, insulin is rapidly removed from the circulation and inactivated by degradation. The liver is the major site of insulin inactivation but in addition nearly all peripheral tissues contain specific insulin-degrading enzymes. The binding of insulin to the plasma membrane receptor is responsible for setting in motion the process that leads to insulin destruction although it is not itself directly responsible for insulin degradation.

The enzymes responsible for insulin degradation are localised within the cell, either soluble (insulin protease) or associated with intracellular membranes (insulin-glutathione transhydrogenase) and insulin must therefore enter the cell before it can be degraded. This is achieved by the endocytosis of areas of plasma membrane rich in insulin-receptor complexes (receptor-mediated endocytosis). Once inside the cell, insulin dissociates from the receptor within the endocytosed vesicles. The two then appear to enter separate pathways since insulin is rapidly degraded while the receptor may be recycled to the plasma membrane. Degradation of insulin follows either from the lysis of the endocytosed vesicles or from their interaction with other membranous structures, since the enzymes responsible for insulin degradation are either soluble or membrane bound.

The degradation of insulin may involve sequential attack by insulin-glutathione transhydrogenase followed by insulin protease. Insulin-glutathione transhydrogenase reductively cleaves the molecule, via the intrachain disulphide bonds, into the separate A and B chains, and insulin protease cleaves the separated chains into smaller peptides. However, since insulin protease can attack the intact molecule and it has a lower K_m and a greater specificity for insulin than insulin-glutathione transhydrogenase it may be the major physiological insulin-degrading activity.

The amount of insulin-degrading activity in the liver changes in response to various physiological situations. In particular, it decreases during starvation and increases following refeeding. Insulin appears to play an important role in these changes since it is one of the factors that induces the synthesis of the degrading enzymes. The mechanisms controlling insulin degradation are of clinical significance since an excessive degradation of the hormone could be a cause of diabetes mellitus. Indeed several diabetic patients have been described who are resistant to subcutaneous insulin, but who have a normal sensitivity to intravenous insulin. In all cases the resistance appeared to be caused by an increased insulin-degrading activity of the subcutaneous adipose tissue.

Further Reading

Baldwin, S.A. & Lienhard, G.E. Glucose Transport Across Plasma Membranes. Facilitated Diffusion Systems. *Trends in Biochemical Sciences* (1981) *6*, 208-211

Bradshaw, R.A. & Niall, H.D. Insulin-related Growth Factors. *Trends in Biochemical Sciences* (1978) *3*, 274-278

Czech, M.P. *et al*. The Insulin Receptor: Structural Features. *Trends in Biochemical Sciences* (1981) *6*, 222-225

Denton, R.M. *et al*. A Partial View of the Mechanism of Insulin Action. *Diabetologia* (1981) *21*, 347-362

Goldfine, I.D. Interaction of Insulin, Polypeptide Hormones, and Growth Factors with Intracellular Membranes. *Biochimica et Biophysica Acta* (1981) *650*, 53-67

Goldstein, B.J. & Livingston, J.N. Insulin Degradation by Insulin Target Cells. *Metabolism* (1981) *30*, 825-835

Jefferson, L.S. Role of Insulin in the Regulation of Protein Synthesis. *Diabetes* (1980) *29*, 487-496

Kahn, C.R. *et al*. Insulin Receptors, Receptor Antibodies and the Mechanism of Insulin Action. *Recent Progress in Hormone Research* (1981) *37*, 477-538

Kahn, C.R. What is the Molecular Basis for the Action of Insulin? *Trends in Biochemical Sciences* (1979) *4*, 263-266

Kiechle, F.L. *et al*. Partial Purification from Rat Adipocyte Plasma Membranes of a Chemical Mediator Which Simulates the Action of Insulin on Pyruvate Dehydrogenase. *Journal of Biological Chemistry* (1981) *256*, 2945-2951

Larner, J. *et al*. Generation by Insulin of a Chemical Mediator that Controls Protein Phosphorylation and Dephosphorylation. *Science* (1979) *206*, 1408-1410

Mortimer, G.E. Mechanisms of Cellular Protein Catabolism. *Nutrition Reviews* (1982) *40*, 1-12

Olefsky, J.M. Insulin Resistance and Insulin Action. *Diabetes* (1981) *30*, 148-162

Seals, J.R. & Czech, M.P. Characterization of Pyruvate Dehydrogenase Activator Released by Adipocyte Plasma Membranes in Response to Insulin. *Journal of Biological Chemistry* (1981) *256*, 2894-2899

Straus, D.S. Effects of Insulin on Cellular Growth and Proliferation. *Life Sciences* (1981) *29*, 2131-2139

5 GLUCAGON

Introduction

The existence of a hyperglycaemic factor in the pancreas was postulated in the 1920s when it was found that certain insulin preparations produced an initial hyperglycaemic response when injected into man. However, it was not until the 1950s that the peptide hormone glucagon was purified and its primary structure determined. The existence of purified glucagon made it possible to study in great detail the chemistry and mechanism of action of this hormone. The most outstanding of these studies was the work by Sutherland and his associates in the 1960s on the mechanism of action of glucagon, which led to the discovery of cyclic AMP and paved the way towards an understanding of polypeptide hormone action in general.

Glucagon is a circulating polypeptide hormone originating mainly from the A-cells of the pancreatic islets but also from similar cells in the gastrointestinal tract. Glucagon is normally detected in plasma and tissues by radioimmunoassay and should therefore be called immunoreactive glucagon (IRG). Initially, some of the antisera used for the assay of glucagon cross-reacted with material (gut glucagon-like immunoreactivity) originating from the gut and artefactually high values for plasma glucagon were obtained. However, the use of antisera specific for pancreatic glucagon has given values of 5-80 pg/ml for the normal peripheral plasma glucagon concentration in man after an overnight fast.

Structure

Human glucagon consists of a single polypeptide chain of 29 amino acids and has a molecular weight of 3,485 daltons. The amino acid sequences of many mammalian glucagons are identical to that of human (Figure 5.1). This high degree of structural conservation may reflect evolutionary constraints composed by the glucagon recognition unit on the surface of the target cell. The amino acid content of glucagon differs

1

His-Ser-Gln-Gly-Thr-Phe-Thr-Ser-Asp-Tyr-Ser
|
Phe-Asp-Gln-Ala-Arg-Arg-Ser-Asp-Leu-Tyr-Lys
|
Val-Gln-Trp-Leu-Met-Asn-Thr

29

Figure 5.1: The Primary Structure of Human Glucagon

significantly from insulin in containing methionine and tryptophan but not cystine, proline or isoleucine. These differences have been exploited in studies on the biosynthesis of the pancreatic hormones using radioactively labelled amino acids. The absence of cystine is particularly important since it means that the molecule cannot form disulphide cross-links to stabilise its structure.

Glucagon is normally present in the circulation at a concentration in the region of 10^{-10} M and at this concentration it is thought to exist largely as a monomer. The monomer has an extended flexible structure containing few stable intramolecular hydrogen bonds and there is therefore little defined secondary structure. In more concentrated solutions the hormone assumes a largely helical conformation stabilised by hydrophobic interactions. A similar secondary structure is thought to be induced in the molecule when it interacts with receptors on the plasma membrane of target tissues. Coincident with the assumption of helicity, glucagon monomers associate into trimers and higher oligomers, leading eventually to the production of an amorphous precipitate. The association of glucagon monomers within the trimer is by means of hydrophobic interactions, with the result that hydrophobic residues become buried within the oligomer. The higher percentage of helical secondary structure of the monomer is retained in the trimer with residues 10-25 occurring as an α-helix and residues 5-9 and 26-29 in a less regular, right-handed helical conformation.

The self-association of monomers probably occurs during secretory granule formation in the A-cell, since the A-granules are usually amorphous in character. Amorphous granules containing trimers and higher oligomers are an effective way of concentrating the glucagon molecule for storage and for making the hormone less available to degradation by proteolytic enzymes.

Synthesis

The synthesis of glucagon, like that of the other islet hormones, occurs on the ribosomes of the rough endoplasmic reticulum. This is followed by its transport to the Golgi complex where it is packaged into the secretory granules. Glucagon is synthesised initially as a precursor molecule with a molecular weight in the region of 9,000. Evidence in favour of the existence of biosynthetic precursors of glucagon (proglucagon and preproglucagon) comes from the detection in serum and pancreatic extracts of peptides larger than glucagon which react with antiglucagon antibodies. In addition, similar proteins have been demonstrated in studies of glucagon biosynthesis in isolated islets using radioactively labelled amino acids.

Conversion of proglucagon into glucagon in the secretory granules appears to be a very slow process requiring several hours and like the conversion of proinsulin may involve the sequential activities of trypsin-like and carboxypeptidase B-like enzymes. In view of the slow rate of conversion of proglucagon into glucagon it is not surprising that serum may contain significant quantities of proglucagon.

The question of the functional significance of proglucagon is as yet unanswered. Glucagon has little defined secondary or tertiary structure and it is unlikely therefore that proglucagon plays a role in ensuring the correct folding of the glucagon polypeptide chain. Since glucagon is an exportable protein it is to be expected that it would be synthesised by way of a precursor molecule with the 'signal sequence' to ensure that it is discharged from the ribosome into the cisternal space of the endoplasmic reticulum. However, this need could be fulfilled by a much smaller molecule. It appears that many small polypeptides are synthesised via large precursors and this may therefore simply be a reflection of some essential size requirement of the biosynthetic machinery. In addition, it is possible that the subsequent post-translational modification of a large single chain primary gene product may give rise to a variety of biologically active substances. In this context the demonstration of glicentin in the A-cell is significant.

Heterogeneity of Circulating Glucagon in Man

When a plasma sample or pancreatic tissue extract is subjected to gel-filtration chromatography and the fractions assayed for IRG using 'glucagon-specific' antiserum, at least four fractions of different molecular weight may be found. These fractions elute at positions corresponding to molecular weights of approximately 2,000, 3,5000, 9,000 and 150,000·daltons. The 2,000 dalton fraction probably represents a degradation product of the 3,500 dalton native glucagon molecule, while the 9,000 dalton fraction is probably a biosynthetic precursor of glucagon. The 150,000 dalton fraction has been called 'big plasma glucagon' and it is present in the γ-globulin fraction of serum. It is widely regarded as glucagon bound to a large molecular weight carrier protein, although the demonstration of its presence in aqueous extracts of islet tissue suggests that it may be a biosynthetic precursor.

The major fractions in pancreatic extracts are the 3,500 dalton (95 per cent) and the 9,000 dalton (5 per cent) components, although all four fractions may be found in plasma. The various molecular weight components of plasma IRG differ in their contribution to the total circulating glucagon immunoreactivity in various physiological and pathological conditions. In healthy subjects plasma IRG is distributed largely between the 150,000 and 3,500 dalton components in highly variable proportions. Patients with chronic renal failure have elevated plasma glucagon levels (500 pg/ml) 60 per cent of which is the 9,000 dalton component. This finding suggests that the 9,000 dalton component (proglucagon) is cleared from the circulation largely by the kidney.

It is now generally recognised that many peptide hormones are heterogeneous in the circulation and that the interconversion and metabolism of these various forms adds additional complexity to the understanding of hormone secretion and action.

Glicentin

Glicentin is a 69-amino acid polypeptide with a molecular weight of 8,128 daltons which contains within its amino acid sequence the full sequence of the glucagon molecule. The protein was originally isolated from the gut but has recently been shown to be present with glucagon in the secretory granules of the human A-cell. Morphological evidence suggests that the two proteins are located in separate regions of the granule, the glucagon being present in the dense core of the granule with the glicentin in the small surrounding area of less electron-dense material. Glucagon and glicentin are co-secreted during granule release.

Glicentin cross-reacts with glucagon antibodies especially those directed against the *N*-terminus, although it has none of the biological activity of glucagon and its function is unknown. Its presence in the A-granules, the fact that it contains the full sequence of glucagon and its lack of biological activity all suggest, however, that it may be a bio-synthetic precursor molecule. The presence of glicentin in the gut and in the pancreas has led to the suggestion that all the glucagon-like poly-peptides of the pancreas and gut share a common biosynthetic pathway and may be derived from a common precursor molecule. Indeed, the concept that post-translational modifications of a single primary gene product can generate a family of structurally related polypeptides is well established in endocrinology.

Glucagon Secretion

The molecular and morphological events associated with glucagon secretion are not as well defined as those of the insulin secretory pro-cess. It has however been established that glucagon is secreted from the A-cell by a process involving margination and exocytosis of the A-granules and that microtubules play a role in margination. In addition, it appears that calcium and cyclic AMP are involved. It is likely there-fore that the secretory process will have a similar molecular basis to that of the B-cell although the mechanism of its regulation may differ.

Table 5.1: Factors which may Regulate Glucagon Secretion

Stimuli	Inhibitors
Hypoglycaemia	Hyperglycaemia
Low fatty acid levels	High fatty acid levels
Most amino acids	Ketone bodies
Adrenaline	Secretin
Noradrenaline	Somatostatin
Acetylcholine	Serotonin
Dopamine	
Gastrin	
Pancreozymin	
Gastric inhibitory polypeptide	
Vasoactive intestinal polypeptide	

Table 5.1 lists those factors which are thought to regulate glucagon secretion. The rate of glucagon secretion from the A-cell is controlled

under physiological conditions largely by the concentration of nutrients in the portal circulation. Glucose, fatty acids and ketone bodies inhibit and amino acids such as arginine, alanine, glycine and leucine stimulate secretion. In addition, exercise and stress may promote glucagon secretion, these effects being mediated at least in part by the autonomic nervous system, which plays an important role in the control of the activity of the A-cell. In man hypoglycaemia (blood glucose < 3mM) increases plasma IRG from basal levels of 5-80 pg/ml to 200-300 pg/ml, while hyperglycaemia suppresses the level by about 50 per cent. Alterations in the rate of secretion affect mainly the amount of the 3,500 dalton material in the circulation, there being parallel but smaller changes in the 9,000 dalton material and relatively little change in the 150,000 dalton material.

Glucagon Metabolism

The half-life of circulating glucagon and proglucagon in man are approximately 5 and 16 minutes respectively. The kidney extracts and metabolises all the various circulating IRG components and is the major site of glucagon metabolism. It is believed that glucagon is filtered by the glomerulus and reabsorbed by the proximal tubules, in which it is degraded. The liver also plays an important role in the metabolism of glucagon removing up to 60 per cent of the total amount of glucagon entering via the portal vein, although it does not extract a significant amount of the other circulating IRG components. This is similar to the pattern of metabolism observed for insulin, in that the kidney metabolises proinsulin, insulin and C-peptide but the liver metabolises only insulin.

It was considered unlikely that enzymes specific for the degradation of glucagon existed, since flexible polypeptides like glucagon are readily degraded in the monomeric state by non-specific proteases. However, specific sites for glucagon degradation, distinct from the glucagon receptor, have been demonstrated in the liver. In addition a hepatic glucagon-degrading enzyme system which removes the three *N*-terminal amino acids has been found. The presence of the 2,000 dalton component in serum also suggests a specific degrading activity.

Physiological Actions of Glucagon

The function of glucagon in man is to ensure that circulating fuel molecules are maintained at a concentration appropriate to the varying needs of the tissues of the body under a variety of physiological conditions. It plays a major role in maintaining the blood glucose concentration and preventing hypoglycaemia in the postabsorptive period, during fasting, starvation and exercise, following the ingestion of a protein rich meal and during the early neonatal period. The prevention of hypoglycaemia under these conditions is necessary to ensure that the glucose-dependent tissues such as the central nervous system are not deprived of their major source of energy since this would result in impaired function of these tissues. In addition, glucagon plays a role in elevating plasma ketone body and/or fatty acid levels during starvation, exercise and acute stress and in raising blood glucose levels following trauma, although, other stress hormones such as adrenaline growth hormone, β-endorphin and corisol also have important actions on blood metabolite levels during these situations. These effects of glucagon are mediated by increases in the mobilisation of fuel molecules from their sites of storage and/or synthesis, since glucagon has no direct effect on peripheral fuel utilisation.

In most situations glucagon is antagonistic to insulin in its actions and it is the relative concentrations of these two hormones that is a major factor in determining whether fuel molecules are mobilised or stored. The normal postabsorptive concentration of insulin in the circulation effectively overrides the action of glucagon and most of the physiological actions of glucagon (Table 5.2) only occur when the insulin concentration falls below the postabsorptive value or when the glucagon concentration increases.

The major site of action of glucagon is the liver, where at physiological concentrations it promotes the rapid mobilisation of glucose stored as glycogen (glycogenolysis), the production of glucose from non-carbohydrate precursors (gluconeogenesis) and the production of ketone bodies from fatty acids (ketogenesis). In addition, it has important catabolic effects on muscle and adipose tissue. In adipose tissue it promotes the mobilisation of fatty acids and glycerol stored as triglyceride (lipolysis). The fatty acids released are used as a fuel by peripheral tissues and provide the substrate for hepatic ketogenesis, while the glycerol is a substrate for hepatic gluconeogenesis. In muscle and liver glucagon also promotes the breakdown of protein (proteolysis) increasing the release of amino acids, many of which are important substrates for hepatic gluconeogenesis.

In the postabsorptive state, during the overnight fast and during prolonged exercise the blood glucose concentration is maintained by glucose released from the liver. Most of this glucose (75-85 per cent)

Table 5.2: Effects of Glucagon on Intermediary Metabolism

Effect	Tissue
(a) Carbohydrate metabolism	
Stimulation of glycogenolysis	Liver
Inhibition of glycogen synthesis	Liver
Stimulation of gluconeogenesis	Liver, Kidney cortex
Inhibition of glycolysis	Liver
(b) Lipid metabolism	
Stimulation of lipolysis	Adipose
Stimulation of ketogenesis	Liver
Inhibition of triglyceride synthesis	Liver
(c) Protein metabolism	
Stimulation of proteolysis	Liver, Muscle

comes from the glucose stored in the liver as glycogen and glucagon is the major factor responsible for its mobilisation under these conditions. This effect of glucagon involves the activation of phosphorylase, the rate-controlling step of glycogenolysis. In addition, glucagon inhibits glycogen synthase, the rate-determining step of glycogen synthesis (glycogenesis) and in this way prevents the futile recycling of glucose back to glycogen. The mechanism of action of glucagon in regulating the activities of these enzymes will be discussed in greater detail in the next section.

There is only a limited amount of glucose stored in the liver as glycogen (100-150g) and as this store becomes depleted, gluconeogenesis assumes an increasingly important role in the maintenance of blood glucose. The time scale of the change from glycogenolysis to gluconeogenesis depends on the rate of peripheral glucose utilisation. The major site of gluconeogenesis is the liver with the kidney cortex assuming a significant role during prolonged starvation.

Glucagon stimulates hepatic gluconeogenesis from lactate, pyruvate, glycerol and the glucogenic amino acids especially alanine. These substrates enter the gluconeogenic pathway at various points and it is likely therefore that glucagon affects the pathway at a step that is common to all the substrates. In addition, the gluconeogenic pathway includes several steps which are common to the glycolytic pathway and stimulation of gluconeogenesis must be accompanied by an inhibition of glycolysis if the futile cycling of common intermediates is to be avoided. Thus, glucagon has been shown to inhibit glycolysis at the level of 6-phosphofructo-1-kinase and pyruvate kinase and to stimulate gluconeogenesis at the level of phosphoenolpyruvate carboxykinase and fructose 1,6-bisphosphatase. Glucagon may also increase the supply of gluconeogenic substrates to the liver by promoting muscle proteolysis and adipose tissue lipolysis and may promote the uptake of amino acids into the liver.   The effect of glucagon on gluconeogenesis is of minor importance during the overnight fast but it becomes physiologically important during prolonged periods of fasting, in prolonged exercise and in early neonatal life.

The maintenance of the blood glucose concentration in all these physiological conditions is essential to ensure that the glucose-dependent tissues such as the central nervous system, the kidney medulla and blood cells receive an adequate supply of glucose. These tissues, unlike other tissues cannot use fatty acids as an energy source, and together they require in   the region of 180g of glucose per 24 hours. The central nervous system is the major glucose consumer (140g) and the maintenance of the proper functioning of this system is essential to survival. In starvation the central nervous system replaces part of its glucose requirement with ketone bodies. This adaptive shift towards ketone body metabolism is in part related to an elevation in the concentration of ketone bodies in the circulation.

The elevation in circulating ketone body concentration is caused by an increased rate of hepatic ketone body production. This in turn is due to the increased availability of ketogenic substrate (fatty acids) to the liver and to an increased metabolism of fatty acids via the ketogenic pathway. Glucagon is in part responsible for both of these changes. It increases the availability of fatty acids to the liver by promoting lipolysis in adipose tissue and it switches liver metabolism away from fatty acid esterification towards oxidation and ketogenesis. The mechanism of action of glucagon in promoting lipolysis is related to the activation of triglyceride lipase, the rate-limiting enzyme of this pathway. The effect of glucagon on ketogenesis has not been defined in detail but it is thought to involve the stimulation of fatty acid uptake into the liver mitochondria.

Molecular Basis of Glucagon Action

Studies on the mode of action of glucagon on the liver have contributed greatly to our understanding of the mechanism of action of polypeptide hormones in general. It is thought that many polypeptide hormones ('primary messengers') do not enter target tissues to produce their physiological effects but rather, they interact with specific receptor sites on the plasma membrane of the cell. This interaction produces changes in the intracellular concentration of so called 'second messengers' which mediate the intracellular effects of the hormone. Second messengers are important regulatory molecules which affect the activity of key enzymes and hence control the physiological activity of the cell.

The initial event in the action of glucagon on the hepatocyte is its rapid binding to specific, high-affinity receptors localised on the plasma membrane of the cell. The binding is saturable and occurs over the physiological range of plasma glucagon concentrations (10^{-10}-10^{-8} M). The apparent maximal binding capacity is of the order of 100,000 glucagon molecules per hepatocyte and at a normal circulating plasma glucagon concentration about 10,000 molecules are bound per cell. These values must be regarded as approximations since it has been shown that the number of receptors on the cell can vary under different physiological and pathological conditions and the glucagon itself can regulate the number of receptors. The binding is reversible and glucagon is released from the receptor unmodified.

It is likely that glucagon bound to its receptor has a well-defined helical secondary structure stabilised by hydrophobic interactions. As circulating glucagon has little secondary structure, the receptor, or its environment, must stabilise or even induce, the structure found in the hormone-receptor complex.

The interaction of glucagon with its receptor on the cell surface stimulates the conversion of ATP into cyclic AMP (the second messenger for glucagon) on the inner surface of the plasma membrane. This is because the glucagon receptor is part of a complex allosteric enzyme system, termed adenylate cyclase (Figure 5.2), which spans the plasma membrane. The adenylate cyclase complex is composed of three interacting units, a discriminator unit (receptor) for hormone binding, a transducer unit which links the receptor to the enzyme unit and possesses a nucleotide regulatory site having high specificity for GTP (G-Protein), and a catalytic unit which converts ATP into cyclic AMP.

The activation of adenylate cyclase by glucagon was thought to occur as a result of a close structural relationship between the receptor

(a) Formation

$$ATP \quad \xrightarrow[\text{Adenylate Cyclase}]{Mg^{2+}} \quad \text{cyclic AMP} \quad + \quad \text{Pyrophosphate}$$

(b) Degradation

$$\text{Cyclic AMP} \quad \xrightarrow[\substack{\text{Cyclic Nucleotide} \\ \text{Phosphodiesterase}}]{Mg^{2+}} \quad \text{5' AMP}$$

Figure 5.2: Cyclic AMP Formation and Degradation

and catalytic units, which allowed a hormone-induced conformational change in the receptor to be relayed directly to the catalytic unit. However, it now appears likely that the receptor and inactive catalytic units are distinct structural entities able to move independently in the plane of the plasma membrane. The interaction of the hormone with the receptor leads to the formation of the hormone-receptor complex which interacts with the G-protein and facilitates the binding of GTP to this protein. This activates the G-protein, enabling it to activate the catalytic unit.

The activation of hepatocyte adenylate cyclase by glucagon and the resultant increase in intracellular cyclic AMP concentration, is the first step in a series of events which lead eventually to the physiological response (Figure 5.3). The major physiological responses of the hepatocyte to glucagon are stimulation of glycogenolysis, gluconeogenesis and ketogenesis and they all appear to be mediated via the cyclic AMP system. The extent to which these processes are activated depends on the intracellular concentration of cyclic AMP which is itself related to the extracellular glucagon concentration. Glycogenolysis is sensitive to a small increase in intracellular cyclic AMP, gluconeogenesis is less sensitive and ketogenesis is least sensitive.

The concentration of cyclic AMP in the cell determines the activity of a cyclic AMP-dependent protein kinase enzyme. This enzyme, in the absence of cyclic AMP, consists of four subunits two of which contain binding sites for cyclic AMP (R-subunits) and two of which are responsible for the catalytic activity of the enzyme (C-subunits). These four subunits are arranged in such a way that the R-subunits mask the active sites of the C-subunits and the complex has no activity towards its

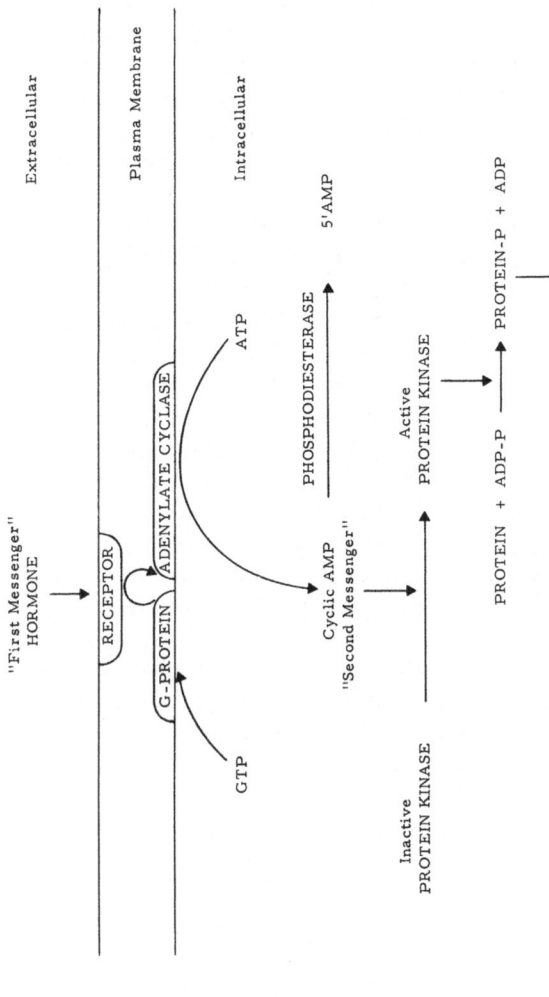

Figure 5.3: The Adenylate Cyclase/Cyclic AMP System

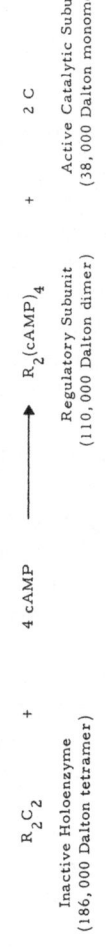

$$R_2C_2 \quad + \quad 4 \text{ cAMP} \quad \longrightarrow \quad R_2(\text{cAMP})_4 \quad + \quad 2 \text{ C}$$

Inactive Holoenzyme   Regulatory Subunit   Active Catalytic Subunits

(186,000 Dalton tetramer)   (110,000 Dalton dimer)   (38,000 Dalton monomers)

Figure 5.4: Activation of Cyclic AMP-dependent Protein Kinase

substrate. In the presence of cyclic AMP the R-subunits bind cyclic AMP and dissociate from the complex unmasking the catalytic sites (Figure 5.4). The extent of dissociation and hence the degree of activity depends on the cyclic AMP concentration. The activated enzyme is responsible for phosphorylating and thereby altering the activity of key regulatory enzymes of the metabolic pathways involved in the physiological response (Figure 5.5).

Some of the events involved in the activation of hepatic glycogenolysis by glucagon are shown in Figure 5.6. This cascade of events involves changes in the catalytic activities of enzymes and results in the progressive amplification of the original glucagon signal. The degree of amplification is so great that one molecule of glucagon can stimulate the release of approximately 3×10^6 molecules of glucose from the liver cell. Approximately half of this amplification occurs at the adenylate cyclase activation step. The cyclic AMP-mediated increase in cyclic AMP-dependent protein kinase activity leads to the phosphorylation and increase in activity of phosphorylase kinase. This enzyme in turn phosphorylates phosphorylase and converts it into a form that is active in the absence of allosteric modification. Phosphorylase is the rate-determining step of glycogenolysis and attacks the 1-4 a-linked glucose units of glycogen liberating the glucose unit as glucose-1-phosphate.

Cyclic AMP-dependent protein kinase also phosphorylates glycogen synthase, the rate-determining step of glycogenesis. In this case however, the phosphorylation of the enzyme converts it into a form which is inactive even in the presence of allosteric factors and as a result glycogenesis is inhibited. The inhibition of glycogenesis at a time when glycogenolysis is stimulated effectively prevents the reincorporation of the liberated glucose back into glycogen and thereby stops the futile cycling of glucose between these two pathways.

The action of glucagon on gluconeogenesis, like its action on glycogenolysis also involves the cyclic AMP-mediated activation of a cyclic AMP-dependent protein kinase. This enzyme phosphorylates pyruvate kinase converting it into an inactive form and effectively blocking the conversion of phosphoenolpyruvate into pyruvate and enabling phosphoenolpyruvate to enter the gluconeogenic pathway. In addition, the protein kinase phosphorylates and inactivates 6-phosphofructo-2-kinase. This enzyme is responsible for the synthesis of fructose 2,6-bisphosphate, a potent allosteric activator of 6-phosphofructo-1-kinase. These alterations effectively inhibit glycolysis and activate gluconeogenesis.

Glucagon is able to switch hepatic fatty acid metabolism away from esterification towards oxidation and ketogenesis. This effect of glucagon

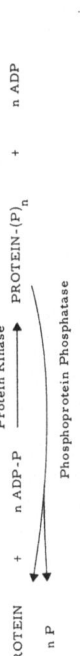

Figure 5.5: Regulation of Protein Phosphorylation

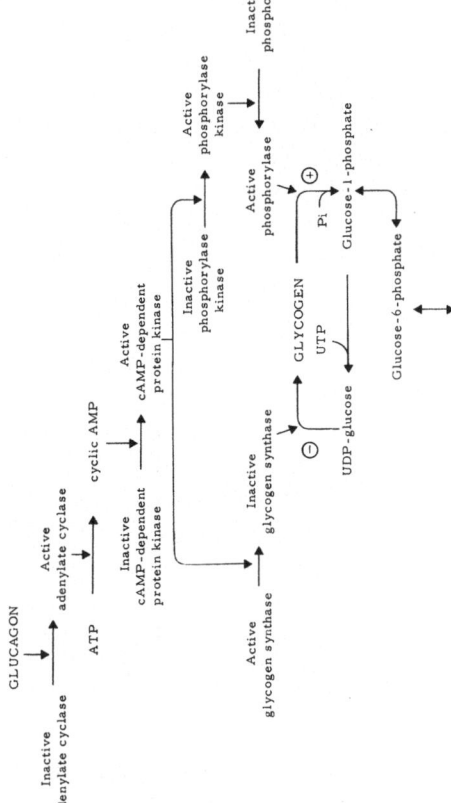

Figure 5.6: The Activation of Hepatic Glycogenolysis by Glucagon. This diagram represents a simplified version of the events involved in the activation of glycogenolysis as the involvement of phosphoprotein phosphatases has been omitted. - reaction inhibited; + reaction stimulated.

is thought to be mediated via the cyclic AMP system and to involve the stimulation of fatty acid uptake into liver mitochondria where they undergo β-oxidation and conversion into ketone bodies. Mitochondria are relatively impermeable to fatty acids and in order to be transported into mitochondria fatty acids have to be converted into their fatty acylcarnitine derivatives (Figure 5.7). This involves the activity of the fatty acyl-CoA carnitine acyltransferase enzymes located in the inner mitochondrial membrane. This enzyme system is rate-limiting for fatty acid transport and it is likely that glucagon's effect on ketogenesis is mediated through the activation of this enzyme system.

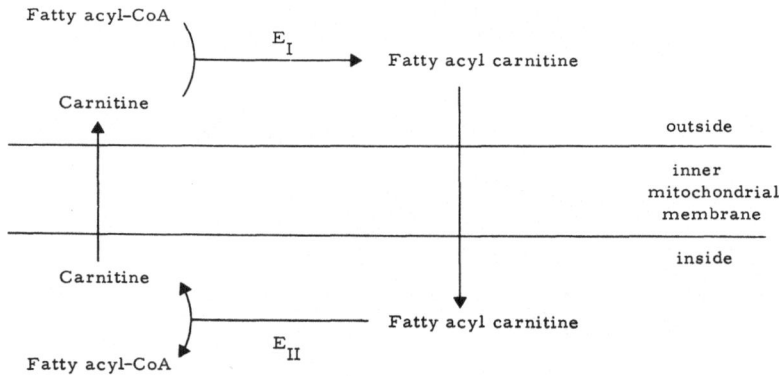

Figure 5.7: Transport of Fatty Acids into Mitochondria. E_I = Fatty acyl carnitine acyl transferase I; E_{II} = Fatty acyl carnitine acyl transferase II.

The effect of glucagon in promoting lipolysis in adipose tissue also appears to be mediated via the cyclic AMP system. Thus, the rate-limiting enzyme of lipolysis, triglyceride lipase, can be activated by phosphorylation. This is achieved by a cyclic AMP-dependent protein kinase and glucagon via cyclic AMP can activate this enzyme.

Thus most, if not all of the effects of glucagon on its target tissue are mediated by cyclic AMP by a mechanism which is outlined in Figure 5.4. This is however an oversimplification of the true picture since the role of phosphoprotein phosphatase enzymes is not shown. These are enzymes which remove the phosphate from phosphorylated protein substrates thereby affecting the degree of phosphorylation and the extent of their functional activity. It is thought that regulation of the activity of these enzymes may exert important controls over the physiological response.

Glucagonoma

Patients with abnormally elevated circulating concentrations of glucagon caused by an A-cell tumour (glucagonoma) have been described. In general, the tumours are of two types: benign, localised A-cell adenomas or slowly metastasising malignant tumours. The benign tumours are generally asymptomatic and are therefore only rarely diagnosed. They produce excessive amounts of glucagon and some of its biosynthetic precursors which are released into the circulation. The resultant hyperglucagonoma is however clinically silent. The malignant tumours produce a hyperglucagonaemia which is more marked than in the benign form (> 1000 μg/ml) and is associated with large amounts of precursor molecules. In addition, such tumours are often associated with hyperglycaemia, although this is usually mild and is rarely associated with ketoacidosis.

Elevated levels of glucagon have been detected in a number of patients who do not appear to have A-cell tumours. Gel-filtration analysis of the glucagon immunoreactivity in these patients has shown that the major component has a molecular weight in the region of 9,000 and probably represents a biosynthetic precursor molecule. The presence of high levels of a glucagon precursor suggests that these patients may have a defect in the enzyme system responsible for converting proglucagon into glucagon. This defect appears to be inherited as an autosomal dominant character.

Glucagon Deficiency

Cases of severe neonatal hypoglycaemia apparently caused by glucagon deficiency have been reported. The circulating glucagon concentration was shown to be low, in spite of the hypoglycaemia, and did not respond to alanine which is normally a potent stimulus of glucagon secretion. The disorder appears to be a lethal one and this may explain the absence of cases of hypoglucagonaemia in adults.

Further Reading

Blundell, T. Conformation and Molecular Biology of Polypeptide Hormones II Glucagon. *Trends in Biochemical Sciences* (1979) *4*, 80-83

Conlon, J.M. The Glucagon-Like Polypeptides — Order out of Chaos? *Diabetologia* (1980) *18*, 85-88

Cooper, D.M.F. Bimodal Regulation of Adenylate Cyclase. *FEBS Letters* (1982) *138*, 157-163

Limbird, L.E. Activation and Attenuation of Adenylate Cyclase. *Biochemical Journal* (1981) *195*, 1-13

Pilkis, S.j. *et al*. Fructose 2,6-bisphosphate: a mediator of hormone action at the fructose-6-phosphate/fructose 1,6-bisphosphate substrate cycle. *Molecular and Cellular Endocrinology* (1982), *25*, 245-266

Schade, D.S., et al. The Role of Glucagon in the Regulation of Plasma Lipids. *Metabolism* (1979), *28*, 874-886.

Unger, R.H. & Orci, L. Glucagon and the A cell. *New England Journal of Medicine* (1981), *304*, 1518-1524; 1575-1580

6   SOMATOSTATIN

Introduction

In the early 1970s while analysing extracts of ovine hypothalamus for a growth hormone releasing factor, Guilleman and co-workers observed that the material from some fractions of the purification sequence inhibited the secretion of growth hormone. The major component with the growth hormone release-inhibiting activity was isolated, characterised and named somatostatin. It is now known that somatostatin is also present in other areas of the central nervous system and in peripheral ganglia and that most somatostatin is in fact located outside the central nervous system, predominantly in the pancreas, stomach and small intestine. The highest concentrations of somatostatin are found in the D-cells of the pancreatic islets and in areas of the hypothalamus. Furthermore, it is now appreciated that somatostatin has many important physiological actions in addition to inhibiting growth hormone secretion (Table 6.1). The major effects of somatostatin appear to be on endocrine tissue function and it may be that some of the non-endocrine actions are secondary to its inhibitory effect on endocrine activity.

Structure

Somatostatin is a cyclic tetradecapeptide with the primary structure shown in Figure 6.1. The molecule has been synthesised by chemical means and the availability of the synthetic molecule led to the rapid development of radioimmunoassay procedures for its measurement and has enabled numerous studies to be performed to investigate its actions. The reduced non-cyclised form of synthetic somatostatin, dihydrosomatostatin, has the full biological activity of the native molecule, both being significantly active at a concentration of 10^{-9}M *in vitro*.

Table 6.1: Biological Activities of Somatostatin

Endocrine — inhibition of secretion of:

(a) Pituitary
      Growth hormone
      Adrenocorticotropin
      Thyrotropin

(b) Pancreatic islets
      Insulin
      Glucagon
      Pancreatic polypeptide

(c) Gastrointestinal tract
      Gastrin
      Pancreozymin
      Secretin
      Vasoactive intestinal peptide
      Gastric inhibitory polypeptide
      Motilin
      Gut glucagon-like immunoreactivity

Non-endocrine — inhibition or reduction of:

(a) Gastrointestinal tract
      Gastric acid secretion
      Pancreatic bicarbonate and enzyme release
      Gastric motility
      Gall bladder contraction

(b) Liver
      Splanchnic blood flow

Synthesis and Storage

Somatostatin is synthesised in the D-cells of the pancreatic islets as a large precursor molecule, prosomatostatin, which is subsequently converted into somatostatin by proteolytic cleavage. The newly formed somatostatin is stored in membrane limited vesicles prior to secretion from the cell.

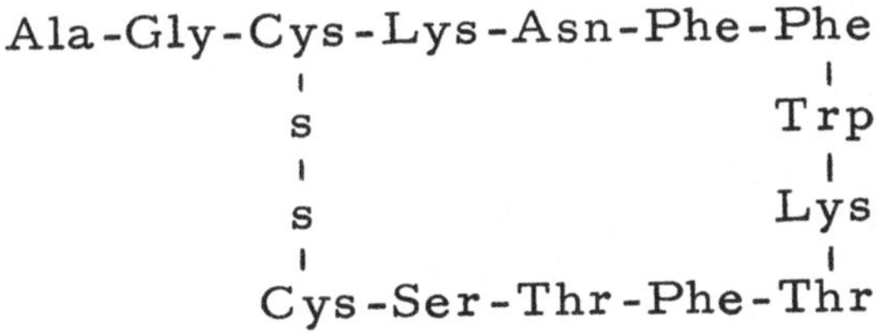

```
Ala-Gly-Cys-Lys-Asn-Phe-Phe
         |                 |
         s                Trp
         |                 |
         s                Lys
         |                 |
      Cys-Ser-Thr-Phe-Thr
```

Figure 6.1: The Primary Structure of Human Somatostatin

Metabolism

Somatostatin has a very short (approx. 1 min) half life in the circulation, being rapidly inactivated by peptidases in the plasma. This, together with the equivocal demonstration of its presence in the systemic circulation, suggests that it is metabolised close to its sites of production.

Secretion

The release of somatostatin from the D-cells of the islet occurs via the process of exocytosis, in which the storage granules containing the hormone fuse with the plasma membrane of the cell, releasing the granule contents into the extracellular space. It appears that both calcium and cyclic AMP may be involved as intracellular messengers controlling the rate of secretion in response to a variety of extracellular stimuli.

There is a remarkable parallelism between the agents that promote the secretion of somatostatin from the D-cell (Table 6.2) and those that stimulate the release of insulin from the B-cell (Table 3.3) since the same metabolites and gastrointestinal tract hormones promote the secretion of both islet hormones. The characteristics of the D-cell secretory response to glucose are also very similar to those of the B-cell to this agent (Chapter 3). Thus there is a sigmoid relationship between the extracellular glucose concentration and the increase in somatostatin secretion, with the threshold glucose concentration being close to 5mM. Moreover, the effect of glucose on secretion is inhibited by mannoheptulose and mimicked by glyceraldehyde and dihydroxyacetone. These observations suggest that the recognition of glucose by the D-cell may involve the metabolism of glucose (Hermansen, 1981).

Table 6.2: Agents Which Affect Somatostatin Secretion

Stimuli

   Glucose
   Arginine, leucine, amino acid mixtures
   Pancreozymin-cholecystokinin
   Gastrin
   Gastric inhibitory polypeptide
   Secretin
   Glucagon

Inhibitors

   Adrenaline
   Diazoxide

Somatostatin has a variety of actions on a number of tissues (Table 6.1) and it has been classified as both a neurotransmitter and a hormone (Luft *et al.*, 1978). The presence of somatostatin-containing vesicles in nerve terminals throughout the brain, spinal cord and peripheral ganglia and the demonstration of its ability to alter neuronal function have led to the suggestion that it may act as a peptidergic neurotransmitter. However, the most characterised action of somatostatin concerns its hormonal activity, although it does not appear to act as a conventional systemic hormone since it has a very short half life in the circulation and has not been equivocally demonstrated in the systemic circulation. The wide distribution of somatostatin producing cells and the observation that it exerts effects only on those tissues where it has been localised, suggest that the peptide might act locally in the places where it is produced and it is therefore appropriate to consider its actions in relation to its sites of production.

Islets of Langerhans

Somatostatin inhibits the secretory activity of both the glucagon producing A-cells and the insulin producing B-cells in the islets and thus may influence the amount of insulin and glucagon released from the islets. It has been suggested that somatostatin may be a more important regulator of A-cell function than of B-cell function because of the closer association of A and D-cells in the islet and because glucagon but not insulin stimulates the release of somatostatin.

Gastrointestinal Tract

Somatostatin appears to act as a local regulator of both the exocrine and endocrine functions of the gastrointestinal tract. Its actions (Table 6.1), result in the reduction of a variety of digestive functions which lead to a decrease in the rate of nutrient entry into the portal circulation. This effect on nutrient uptake is probably physiologically important in nutrient homeostasis because nutrients such as glucose and amino acids, and the gastrointestinal hormones pancreozymin and secretin, all stimulate the secretion of somatostatin. Accordingly, gastrointestinal hormones released at the start of nutrient absorption, together with the nutrients themselves, determine the rate of somatostatin secretion and thereby exert negative feedback control over the rate of nutrient entry. Such a system would operate to ensure that the flux of exogenous nutrients into the circulation was within acceptable limits.

Mechanism of Action

It is generally assumed that, as with other polypeptide hormones, the initial step in the mechanism of action of somatostatin involves its binding to membrane receptors on the external surface of target cells. Such receptors have been identified in the anterior pituitary. In addition, a soluble somatostatin binding protein has been isolated from various tissues although its function is unknown. It seems likely that somatostatin acts on the pancreatic B-cell by altering the cell's metabolism of calcium, since it has been shown to impair the uptake of calcium by B-cells. In addition, its effect on the B-cell is reversed by increasing extracellular concentration of calcium or by using a calcium ionophore to increase uptake of calcium (Pace & Tarvin, 1981). Calcium plays an important role in the regulation of most, if not all of the processes that somatostatin is known to affect and it is possible, therefore, that somatostatin may have a single mechanism of action involving modification of calcium metabolism in all its target tissues.

Physiological Role

The rate of disposal of ingested nutrients depends on their rate of uptake into the circulation and their rate of removal from the circulation into tissues for metabolic transformation. Co-ordinated regulation of both these processes is necessary to ensure that the rate of nutrient entry is balanced by the rate of nutrient removal so that major alterations in blood nutrient levels are prevented. Evidence has accumulated which suggests that somatostatin may play a role in regulating both nutrient entry and nutrient disposal. Thus, somatostatin is located in the D-cells of the gastrointestinal tract and the endocrine pancreas. Its secretion is regulated both by metabolic and hormonal factors and it can affect the function of both the endocrine pancreas and the gastrointestinal tract. Accordingly, somatostatin may be a major factor in integrating information from ongoing digestive and metabolic events and providing a fine tuning of the rate at which certain nutrients enter the circulation in balance with their rate of removal (Schusdziarra, 1980).

Somatostatinoma

Somatostatin-containing D-cell tumours of the endocrine pancreas have been identified in a number of patients (Unger, 1977). The patients had low plasma levels of insulin and glucagon and exhibited mild diabetes with an abnormal glucose tolerance but without severe hyperglycaemia or ketosis. In addition, achlorhydria, steatorrhea and gall bladder disease were noted in the patients, possibly related to the inhibitory actions of somatostatin on gastric acid secretion, pancreatic enzyme secretion and gall bladder contraction. These observations lend support to the idea that somatostatin may play a physiologically important role in regulating the activity of both the endocrine pancreas and the gastrointestinal tract.

Current and Potential Uses

Somatostatin has been used as an experimental tool in man to induce a deficiency of certain hormones in attempts to assess their physiological or pathological importance. For example, it has proved useful in establishing a physiological role for glucagon in sustaining normoglycaemia in man. Thus infusion of somatostatin into volunteers was associated with a 30 to 50 per cent fall in blood glucose levels as the result of a parallel fall in hepatic production of glucose. This effect was mediated by suppression of glucagon secretion since somatostatin has no direct effect on glucose metabolism. In addition, it has been used to establish the importance of glucagon in exaggerating the hyperglycaemia and other metabolic consequences of insulin deficiency in human diabetes. Thus, infusion of somatostatin for 18 hours after the acute withdrawal of insulin from type 1 diabetic subjects, reduced the hyperglycaemia and prevented the development of ketoacidosis. This action, which has been attributed to suppression of glucagon secretion, has raised the possibility that somatostatin might prove useful as an adjunct to insulin in the management of diabetes.

The ability of somatostatin to inhibit a number of physiological functions and the relative ease with which such a small molecule can be synthesised chemically have led to attempts to find therapeutic uses for the hormone. However, the short biological half-life and the broad specificity of somatostatin have limited its usefulness as a therapeutic agent. In an effort to produce a substance that would have a longer duration of activity, a greater biological potency and differing specificities, a variety of analogues of somatostatin have been synthesised. In particular, a search has been made for an analogue with A-cell inhibitory specificity which could be used to reduce the relative or absolute hyperglucagonaemia seen in diabetes and might therefore be expected to reduce the severity of the metabolic complications of the disease. The analogue Cys^{14}-somatostatin has been shown to be more effective in suppressing glucagon secretion than insulin secretion, while Trp^8-Cys^{14}-somatostatin and Trp^8-somatostatin appear to enhance the effectiveness of insulin therapy in the treatment of hyperglycaemia.

References and Further Reading

Efendic, S. Somatostatin. *Advances in Metabolic Disorders* (1978) *9*, 367-424

Gerich, J.E. & Patton, G.S. Somatostatin. *Medical Clinics of North America* (1978) *62*, 375-392

Hermansen, K. Pancreatic D-Cell Recognition of D-Glucose. *Diabetes* (1981) *30*, 203-210

Luft, R. *et al*. Somatostatin – Both Hormone and Neurotransmitter? *Diabetologia* (1978), *14*, 1-13

Pace, C.S. & Tarvin, J.T. Somatostatin: Mechanism of Action in Pancreatic Islet β-cells. *Diabetes* (1981) *30*, 836-842

Schusdziarra, V. Somatostatin – A Regulatory Modulator Connecting Nutrient Entry and Metabolism. *Hormone and Metabolic Research* (1980) *12*, 563-577

Unger, R.H. Somatostatinoma. *New England Journal of Medicine* (1977) *296*, 998-1000

7 PANCREATIC POLYPEPTIDE

Introduction

A 36 amino acid polypeptide with hormonal properties was isolated in the early 1970s from the pancreas of chickens and subsequently from the pancreas of a number of mammalian species including man. This protein, pancreatic polypeptide, is synthesised and stored in a distinct islet cell type, the pancreatic polypeptide (PP) cell. The structure of the protein has been determined in several species, including man (Figure 7.1) and sensitive radioimmunoassays have been developed which allow its plasma concentration to be determined. In addition, several actions have been ascribed to the protein, and factors involved in the regulation of its secretion have been determined. However, in spite of intensive study, the physiological function of pancreatic polypeptide has not yet been ascertained. In this chapter I shall review the information currently available concerning human pancreatic polypeptide, in the hope that it may provide a basis for the understanding of subsequent studies.

1
Ala-Pro-Leu-Glu-Pro-Val-Tyr-Pro-Gly-Asp-Asn-Ala
                                                  |
Leu-Asp-Ala-Ala-Tyr-Gln-Ala-Met-Gln-Glu-Pro-Thr
|
Arg-Arg-Tyr-Ile-Asn-Met-Leu-Thr-Arg-Pro-Arg-Tyr-NH$_2$
                                                      36

Figure 7.1: The Primary Structure of Human Pancreatic Polypeptide

Structure

Pancreatic polypeptide is a straight chain polypeptide of 36 amino acids with a molecular weight in the region of 4,300 daltons. The proteins isolated from several mammalian species differ by only 1 or 2 residues, although less than half of the residues of mammalian pancreatic polypeptide are conserved in the primary structure of the avian protein. Nevertheless, mammalian and avian pancreatic polypeptide appear to have similar secondary structures in dilute, aqueous solution. The primary structure of human pancreatic polypeptide is shown in Figure 7.1. The protein has a free *N*-terminus and a *C*-terminal tyrosine amide. The blocked *C*-terminus is a feature common to all pancreatic polypeptides so far sequenced. Crystallographic studies suggest that the molecule has a relatively stable tertiary structure with a polyproline helix involving residues 2, 5 and 8 close-packed against an *a*-helical region of residues 14-32 (Figure 7.2). The molecule forms dimers and further polymerises in the presence of zinc ions. It appears that the *C*-terminal portion of the molecule is important for its biological activity since removal of the *C*-terminal tyrosine amide abolishes its activity while a *C*-terminal hexapeptide of pancreatic polypeptide is able to mimic the actions of the whole molecule.

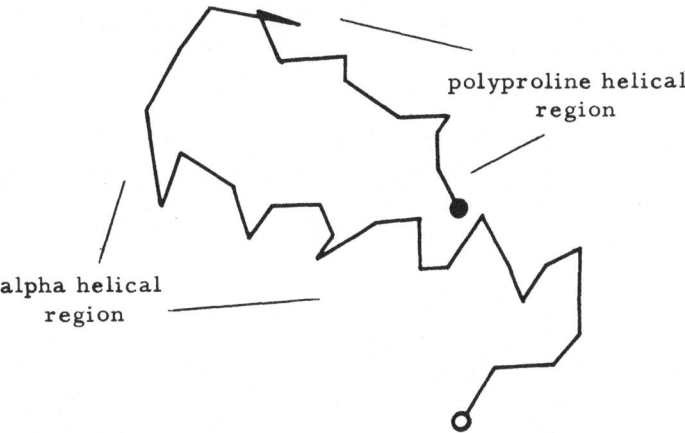

polyproline helical region

alpha helical region

Figure 7.2: Possible Conformation of Human Pancreatic Polypeptide in Solution. ● = *N*-terminus; ○ = *C*-terminus.

Synthesis and Storage

Human pancreatic polypeptide is synthesised in the PP-cells which are largely confined to islets located in the posterior portion of the head of the pancreas. Indeed, the PP-cells are the most frequent endocrine cell type in this region of the pancreas. Pancreatic polypeptide-containing cells have also been located in parts of the gastrointestinal tract. However, the pancreas appears to be the major source of the pancreatic polypeptide which circulates in the basal state, since pancreatectomy reduces the basal level to less than 20 per cent of normal. The PP-cells in man are generally smaller than the other islet cells and are located at the periphery of the islet. The number of PP-cells in the pancreas appears to increase with age and there is a marked hyperplasia of the cells in type 1 diabetes of long duration.

The relatively small size of pancreatic polypeptide suggests that it may be synthesised by post-translational modification of a larger precursor molecule. Recently a 9,000 dalton molecular weight biosynthetic precursor containing the pancreatic polypeptide sequence at its *N*-terminal end, has been isolated from canine pancreas, confirming this suggestion. *In vivo* processing of this precursor gave rise to two proteins, pancreatic polypeptide with a short *C*-terminal extension, and a small unidentified protein of molecular weight 2,500-3,000. The *C*-terminal extension on the pancreatic polypeptide sequence may play a role in the final production of the amidated protein since amidation of the *C*-terminus is a post-translational modification found in many secretory peptides.

Pancreatic polypeptide is stored in the PP-cells of the pancreas in cytoplasmic storage granules of moderate to high electron density. The granules are generally smaller than those of the other islet cells and the granule membrane is closely applied to the dense core.

Secretion

The secretion of pancreatic polypeptide from the PP-cell appears to be largely under the control of the autonomic nervous system, although blood nutrient and hormone levels may also play an important role (Table 7.1). Cholinergic stimulation via the vagus provides the major signal for secretion. In addition, variations in the adrenergic tone may modulate the response, β-receptor agonists promoting secretion and a-receptor agonists inhibiting secretion.

The major physiological situation associated with the secretion of pancreatic polypeptide is the ingestion of food. The magnitude and duration of the response depends on the nutrient content of the food, although the response itself is probably relatively non-specific since an increase in plasma pancreatic polypeptide accompanies the ingestion of a protein-rich meal, a low calorie, fibre-rich meal and even follows the drinking of water. This latter observation suggests that the response may be related in part to gastric distension and the consequent vagal stimulation. The ingestion of a protein-rich meal induces a biphasic secretory response which lasts for several hours, with peak plasma concentrations being reached at 5 and 180 min. The first phase of the response appears to involve vagal control, while the second phase may involve vagal, metabolic and hormonal control.

Changes in the concentration of glucose in the blood have an effect on plasma pancreatic polypeptide levels since hypoglycaemia is associated with increased levels of pancreatic polypeptide while hyperglycaemia reduces the level. These effects do not appear to be mediated directly by glucose since it has only a small effect on the rate of pancreatic polypeptide secretion, and they must therefore reflect the action of neural or endocrine factors released in response to the hyperglycaemia or hypoglycaemia.

Table 7.1: Factors which Affect Pancreatic Polypeptide Secretion

Factor	Effect on secretion
Amino acids (arginine)	Stimulates
Glucose	Inhibits
Somatostatin	Inhibits
Acetylcholine	Stimulates
β-Adrenergic agonists	Stimulate
α-Adrenergic agonists	Inhibit

Plasma Concentration

The normal basal plasma concentration of pancreatic polypeptide in young adults is in the region of 30-70 pg/ml and this increases progressively with age, reaching 150-300 pg/ml by the age of 70. The concentration in women tends to be lower than that in men although the values for age matched groups are not significantly different. The half-life of pancreatic polypeptide in the plasma is thought to be between 5 and 10 minutes. However, the protein does not appear to be metabolised by tissues and it is removed from the circulation in its active form by the kidney.

Alterations in the basal plasma concentration of pancreatic poly-peptide have been observed in a number of pathological conditions, although the significance of these observations remains to be determined. Thus, obesity has been shown to be associated with a 50 per cent fall in plasma pancreatic polypeptide concentrations while diabetes appears to be associated with an increased concentration. The extent of the increase in diabetes appeared to correlate with the severity of diabetes, since type 1 diabetics had a mean value of 170 pg/ml while type 2 diabetics had a mean value of 113 pg/ml compared with the control non-diabetic value of 85 pg/ml.

Pancreatic polypeptide secreting tumours have been found in man, and their detection depends largely on the demonstration of a plasma pancreatic polypeptide concentration at least four times that of age-matched controls. In spite of the high levels of circulating pancreatic polypeptide, patients with this type of tumour do not present with a clear-cut clinical picture and they may even be asymptomatic. This lack of obvious symptoms, associated with high circulating levels of pancreatic polypeptide, makes it difficult to ascribe physiologically important actions to the protein.

Actions and Physiological Role

Pancreatic polypeptide has been shown to affect several pancreatic and gastrointestinal functions, although the physiological significance of these observations is not clear, as many of the effects appear to require pharmacological concentrations of the protein (Table 7.2).

Table 7.2: Actions of Pancreatic Polypeptide

(a)   Endocrine pancreas

        Inhibits insulin secretion
        Inhibits somatostatin secretion

(b)   Gastrointestinal tract

        Inhibits pancreatic zymogen secretion
        Decreases gall bladder contractility
        Reduces gastrointestinal motility
        Inhibits gastric acid secretion

The localisation of PP-cells within the islets of Langerhans suggests that they may exert paracrine effects on the other endocrine cells of the islets. In support of this suggestion it has been shown that pancreatic polypeptide inhibits the secretion of both insulin and somatostatin, although it does not appear to affect the release of glucagon. In addition to affecting the endocrine pancreas, pancreatic polypeptide also affects the exocrine pancreas, inhibiting both basal and stimulated trypsin and bicarbonate secretion. This latter effect is one of a number of actions of pancreatic polypeptide which oppose those of cholecystokinin. These also include the ability of pancreatic polypeptide to decrease gall bladder contractility, reduce gastrointestinal motility and inhibit gastric acid secretion. Thus it appears that the major physiological function of pancreatic polypeptide is to slow down the digestive process. This may be important in helping to ensure that blood nutrient concentrations do not rise too abruptly following meals.

The possible role of pancreatic polypeptide as a regulator of digestive function and the observation that the plasma concentration of the protein is reduced in obesity, have led to the suggestion that pancreatic polypeptide may somehow be involved in the regulation of satiety. This interesting suggestion must however await the results of further studies.

Pancreatic polypeptide is the fourth polypeptide found to be synthesised and stored in the endocrine pancreas. It is released in large amounts into the blood following the ingestion of a meal and it has effects on tissues outside the pancreas. Whilst the physiological significance of these effects are unclear there can be little doubt that pancreatic polypeptide should be classified as a polypeptide hormone. In view of the uncertainty regarding its physiological role and the fact that patients with PP-cell tumours do not show any obvious clinical symptoms that can be related to the very high circulating levels of pancreatic polypeptide, this hormone will not be considered in any further detail.

Further Reading

Blundell, T. & Wood, S. The Conformation, Flexibility, and Dynamics of Polypeptide Hormones. *Annual Review of Biochemistry* (1982), *51*, 123-154

Floyd, J.C. *et al*. A Newly Recognised Pancreatic Polypeptide: Plasma Levels in Health and Disease. *Recent Progress in Hormone Research* (1977) *33*, 519-70

Lantigua, R.A. *et al*. Adrenergic Modulation of Pancreatic Polypeptide Secretion. *Metabolism* (1980) *29*, 787-792

Orci, L. *et al*. Pancreatic polypeptide-rich Regions in Human Pancreas. *Lancet* (1978) 1200-1201

8 THE FUNCTION OF THE ISLETS OF LANGERHANS

Introduction

The islets of Langerhans play an essential role in regulating nutrient homeostasis in man. They function to ensure that despite wide fluctuations in the supply of certain nutrients in the diet, the concentration of these nutrients in the blood is maintained at a level appropriate to the varying and competing needs of the tissues of the body. This is achieved largely by alterations in the rate of secretion of the islet cell hormones into the circulation. These hormones are released in response to changes in the concentration of metabolites and other hormones in the circulation and in response to the activity of the sympathetic and parasympathetic nervous systems. The relative amounts of insulin and glucagon in the circulation, determined largely by their rate of secretion, is an important factor in regulating the rate of flux of metabolites into and out of tissues such as liver, muscle and adipose tissue. Pancreatic polypeptide and somatostatin do not appear to have any direct effect on the metabolism of these tissues, and these hormones may play a role in determining the rate of entry of nutrients into the circulation from the gastrointestinal tract.

Since it is the relative rates of secretion of the islet hormones that largely determine their plasma concentration and hence their action, it is likely that the secretory activities of the various islet cells are coordinated to ensure that the various hormone levels are appropriate to the particular needs of the organism. Evidence in favour of such coordinated activity has been provided by the demonstration of gap-junctions between the various islet cell types which would allow intercellular communication. In addition, it appears that there are paracrine effects between the various islet cell types since the secretory products of one cell affect other cell types (Table 8.1).

The major role of the nutrient homeostasis sytem is to ensure that blood metabolite levels return relatively rapidly to the post-prandial level following the ingestion of a meal, and that this level is maintained during the ensuing fast. Should the fast extend into a period of starvation then metabolite levels must be maintained for as long as possible

116

Table 8.1: Effects of Islet Hormones on the Secretory Activities of Islet Cells

Hormone	Secretory activity			
	Insulin secretion	Glucagon secretion	Somatostatin secretion	Pancreatic polypeptide secretion
Insulin	-	Inhibits	-	-
Glucagon	Stimulates	-	Stimulates	-
Somatostatin	Inhibits	Inhibits	-	Inhibits
Pancreatic polypeptide	Inhibits	-	Inhibits	-

to ensure survival. In addition, there are a number of physiological situations in which blood metabolites need to be maintained above the normal fasting level (Table 8.2). Thus, during exercise and stress blood metabolite levels are increased, while the metabolic response to pregnancy also involves increased blood metabolite levels. Derangements of this homeostatic mechanism, such as occur in diabetes mellitus, have profound effects on blood metabolite levels. In order to understand the acute metabolic derangements of diabetes it is necessary to understand how the islets of Langerhans function normally to regulate blood metabolite levels under a variety of different physiological situations, and to consider the roles of the liver, muscle and adipose tissue in these situations.

The Role of the Liver in Nutrient Homeostasis

The liver plays a major role in the regulation of nutrient homeostasis in man. Depending on the nature and quantity of the nutrients in the diet, the liver can either synthesise glycogen, lipids and protein, and thereby store nutrients, or it can degrade these materials to provide glucose and ketone bodies to serve as fuels for other tissues. These metabolic activities are regulated by the concentration of nutrients in the blood perfusing the liver and by hormones such as insulin, glucagon, glucocorticoids and catecholamines which act directly on the liver to influence its choice of metabolic alternatives. The major effects of insulin and glucagon on liver metabolism are shown in Tables 4.5 and 5.2.

Table 8.2: Blood Nutrient Levels under Various Physiological Conditions

| Condition | Nutrient | | | |
	Glucose (mM)	Fatty acids (mM)	Ketone bodies (mM)	Amino acids (mM)
Basal (postabsorptive)	4-5	0.4-0.8	0.02-0.5	1.2-2.0
Starvation (48 hours)	3-4	0.8-1.2	0.6-1.4	1.4-2.2
Starvation (1 week)	3-4	1.0-2.0	2-10	2.6-3.4
Exercise (sprint)	4-5	0.4-0.8	0.02-0.5	1.2-2.0
Exercise (prolonged)	3-4	0.8-1.6	0.02-0.5	1.4-2.2
Stress (acute)	5-7	0.8-1.6	0.02-0.5	1.2-2.0
Stress (chronic)	5-6	1.0-2.0	0.5-5	2.0-3.0
Pregnancy	3-5	0.5-0.9	0.02-0.5	1.0-1.4

The Role of Muscle in Nutrient Homeostasis

Muscle plays an important role in the maintenance of nutrient homeostasis in man. In times of nutrient excess it can remove glucose and amino acids from the circulation and store them as glycogen and protein respectively. In times of nutrient need it can break down intracellular proteins to provide amino acids for gluconeogenesis and it can switch from glucose utilisation to fatty acid utilisation, and thereby spare glucose for the glucose-dependent tissues. These changes in muscle metabolism are controlled by the concentration of nutrients and hormones in the circulation.

Insulin is essential for the normal growth and development of muscle and it also plays an important role in the regulation of muscle metabolism since it promotes the uptake and utilisation or storage of metabolites such as glucose and amino acids in muscle cells. Glucagon has important effects on muscle protein metabolism although it is unlikely that it affects muscle glucose metabolism. The major effects of insulin and glucagon on muscle metabolism are listed in Tables 4.5 and 5.2.

The Role of Adipose Tissue in Nutrient Homeostasis

Adipose tissue is highly specialised for the storage of triacylglycerols (triglycerides) the major fuel reserve in man (Table 8.3). The triacyl-glycerols stored in adipose tissue are derived exogenously from the diet or are produced endogenously in the liver. In addition, adipose tissue can convert excess blood glucose into triacylglycerols for storage. The storage of triacylglycerols by adipose tissue is regulated by a variety of hormones including insulin and glucagon and is dependent on glucose metabolism by the tissue. Insulin regulates triacylglycerol storage by promoting both the uptake of glucose and the synthesis of fatty acids and their conversion into triglycerides (esterification) while glucagon promotes the breakdown of triglycerides (lipolysis) to release fatty acids and glycerol. Low levels of insulin in the circulation, with or without high levels of glucagon, such as occur physiologically in star-vation and stress, are invariably associated with the mobilisation of lipid from storage and with elevated levels of fatty acids in the circulation. The major effects of insulin and glucagon on adipose tissue metabolism are shown in Tables 4.5 and 5.2.

Feeding

Following the ingestion of a meal, the products of its digestion are absorbed and there is an increase in the concentration of nutrients in the circulation and in particular in the vessels (vascular and lymphatic) draining the gastrointestinal tract. The nutrient homeostatic mechanism responds to this increase to ensure that the level of nutrients returns as quickly as possible to the fasting level. The nature of this response depends in part on the profile of the nutrient changes and this in turn will depend on the contents of the meal.

The absorption of a nutritionally balanced meal (i.e. one consisting of protein, carbohydrate and lipid) or a carbohydrate-rich meal, results in the stimulation of insulin, somatostatin and pancreatic polypeptide secretion and inhibition of glucagon release. The signal for these changes is partly metabolic, due to the increase in blood metabolite levels, and partly hormonal, resulting from the release of hormones such as secretin and pancreozymin into the circulation by the gastro-intestinal tract. In addition, there may be an input from the sympa-thetic and parasympathetic nervous systems.

The elevated levels of pancreatic polypeptide in the circulation and locally released somatostatin may control the rate of nutrient entry from the gastrointestinal tract into the circulation and the elevated levels of insulin ensure that liver and peripheral tissues such as muscle and adipose tissue remove nutrients from the circulation and either utilise them or store them. In this way there are minimal disturbances to blood nutrient levels and they fall rapidly to the pre-absorptive level.

The ingestion of a protein-rich meal leads to an increase in the concentration of amino acids in the portal circulation and to the stimulation of insulin, somatostatin, pancreatic polypeptide and glucagon secretion. Somatostatin and pancreatic polypeptide may control the rate of entry of amino acids into the circulation, and insulin promotes their uptake into peripheral tissues such as muscle and their incorporation into tissue protein. In this way the concentration of amino acids in the circulation rapidly falls to the preabsorptive value. A protein-rich meal is low in carbohydrate and blood glucose levels are maintained during and following the ingestion of such a meal by hepatic glucose production. The increase in circulating insulin concentration induced by amino acid ingestion is usually not sufficient to increase peripheral glucose utilisation but it is sufficient to reduce hepatic glucose production. If this action of insulin was unopposed, hypoglycaemia would follow the ingestion of a protein-rich meal. Hypoglycaemia does not occur however as amino acids in the portal circulation stimulate the secretion of glucagon. The concentration of glucagon in the portal vein is thus sufficient to counteract any effect of insulin on hepatic glucose production and the blood glucose concentration is maintained. In this situation glucose homeostasis is maintained when both the insulin and glucagon concentrations are increased. In most other situations however, these hormone concentrations change in opposite directions.

Fasting and Starvation

Blood nutrient levels are maintained during the overnight fast by mechanisms which ensure that their rate of utilisation is balanced by their rate of release from storage or their rate of formation from available metabolites. The rates of these processes are determined largely by the relative amounts of insulin and glucagon in the blood (i.e. by the insulin/glucagon ratio). During the overnight fast there is a gradual increase in the glucagon level and possibly a small fall in the insulin. These changes are reflections of alterations in the rate of hormone release from islet cells controlled by a gradual decrease in the blood glucose concentration.

The tissues of the body consume in the region of 100g of glucose during the overnight fast (8 p.m. - 8 a.m.). There is only approximately

10g of glucose present in the blood and extracellular fluid at the beginning of the fast and glucose has therefore to be mobilised from storage to meet this demand. The major reserve of glucose is liver glycogen and the decreasing insulin/glucagon ratio promotes the mobilisation of this store. Since the liver only contains in the region of 100g of glycogen, this reserve of glucose would become depleted towards the end of the overnight fast. However, the decreasing insulin/glucagon ratio switches the liver metabolism towards gluconeogenesis and lactate, glycerol and certain amino acids are converted into glucose to maintain blood glucose levels. In addition, the decreasing insulin/glucagon ratio stimulates the release of fatty acids and glycerol from adipose tissue. The fatty acids are used by tissues such as heart and skeletal muscle, thereby sparing glucose for the glucose dependent tissues such as brain, and the glycerol can be used by liver as a substrate for gluconeogenesis.

Should the fast extend for a more prolonged period or should starvation ensue, then more dramatic metabolic adaptations are necessary to ensure that the nutrient requirements of all the essential tissues of the body are met. The major requirement is that the tissues which are totally dependent on glucose should be supplied with glucose, while other tissues should use alternative energy sources such as fatty acids and ketone bodies if possible. Since there is no dietary intake of nutrients during starvation all tissues have to have their needs satisfied from nutrients stored in the body (Table 8.3). The major stores of nutrients are adipose tissue triglyceride, which when hydrolysed releases fatty acids and glycerol, and muscle protein, which releases amino acids. The controlled mobilisation of these reserves is largely under the influence of the insulin/glucagon ratio, although hormones such as growth hormone and cortisol also play an important role.

Table 8.3: Nutrient Stores of a Normal 70 kg Male at the Beginning of a Fast

Type of store	Weight of store (kg)	Energy content of store (kcal)
Triacylglycerol (adipose tissue)	15	100,000
Protein (mainly muscle)	6	25,000
Glycogen (muscle and liver)	0.25	1,000
Circulating nutrients	0.025	100

The mobilisation of adipose tissue triglyceride, under the influence of glucagon and growth hormone, elevates the blood fatty acid level and enables fatty acids to satisfy the major portion of the energy requirement of heart and skeletal muscle. In addition, the increase in blood fatty acid level coupled to the decreasing insulin/glucagon ratio switches liver metabolism to ketone body production (ketogenesis). The blood level of these water-soluble metabolites gradually increases as starvation proceeds and reaches a plateau of about 10mM after a week. The level rarely exceeds this value as the rate of production is under the control of the insulin/glucagon ratio. The increased availability of ketone bodies in the circulation enables the brain to switch from glucose to ketone bodies for some of its energy requirement and the brain glucose requirements drop from 140g/24 h at the beginning of starvation to 40g/24 h after 1 week. This adaptation means that the liver and kidney cortex need only produce 80g of glucose per 24 hours to satisfy the needs of the glucose-dependent tissues. In addition, since muscle protein is the major source of the substrates for gluconeogenesis in these tissues the adaptation effectively conserves the limited reserves of muscle protein and allows them to be used to maintain blood glucose for a longer period. Thus the metabolic adaptations to starvation ensure that nutrient reserves are mobilised and utilised as efficiently as possible enabling periods of starvation lasting up to 60 days to be endured without irreversible damage to tissues.

Exercise

During periods of exercise lasting for more than a few minutes, there is an increasing need for adjustments to be made to the nutrient homeostatic mechanism. This is because the increased energy requirement of heart muscle has to be met largely from nutrients in the blood. In addition, as exercise proceeds skeletal muscle becomes increasingly dependent on blood nutrients for its energy requirements, since muscle glycogen reserves become progressively depleted. Thus there must be a progressive adaptation of the homeostatic mechanism to ensure that the extra energy demands of heart and skeletal muscle are met, without blood nutrient levels falling to a level where central nervous system function becomes impaired. Clearly therefore there is a need to progressively mobilise nutrients such as glucose and fatty acids from storage. In addition, since there is only a limited storage of glucose as glycogen in the liver, muscle cells must gradually adapt to utilise fatty acids, thereby sparing glucose for those tissues with an absolute requirement for glucose. The time scale for the changes in muscle metabolism with exercise is shown in Figure 8.1. Part of the response of the nutrient homeostatic mechanism to exercise involves the endocrine pancreas since there is a progressive need for mobilisation of stored nutrient and this is favoured by a progressive fall in blood insulin levels and a rise in glucagon levels. The time scale of these changes is shown in Figure 8.2 and they are related to effects of catecholamines, released during exercise, on the A- and B-cells which result in the inhibition of insulin secretion and stimulation of glucagon secretion. In addition, other neural and maybe metabolite effects may be involved.

Pregnancy

The metabolic response to pregnancy involves a sequence of changes in maternal metabolism which initially represents a preparatory accumulation of maternal nutrients and then the subsequent growth of the feto-placental unit. Thus, the first two trimesters of a normal pregnancy are characterised metabolically by an increased intake of nutrients by the mother and the storage of excess nutrients principally as maternal adipose tissue. During the final trimester, however, there is relatively little maternal storage of nutrients. This is the period when the placenta and fetus undergo their dramatic growth and as both are totally dependent on maternal nutrients for their growth, there is little left for storage.

The maternal endocrine pancreas plays an essential role in the metabolic adaptation to pregnancy. This involves a gradual increase in the sensitivity of the pancreatic B-cell to secretory stimuli such as glucose. Thus, early in pregnancy when food intake begins to increase, peripheral tissues have a normal (i.e. pre-pregnancy) sensitivity to insulin and insulin released in response to meals ensures that excess nutrients are converted into fat for storage. However, as pregnancy proceeds there is a progressive increase in the plasma level of the placental hormones, progesterone, oestrogen and placental lactogen. These hormones function to ensure that maternal nutrients are available to the fetus rather than for storage in maternal tissues. They achieve this, in part, by reducing the sensitivity of peripheral tissues to insulin. There is normally, however, an adaptive change in the insulin secretory mechanism of the B-cell to this situation, which enables more insulin to be released in response to a given stimulus. Thus, by releasing more insulin the endo-

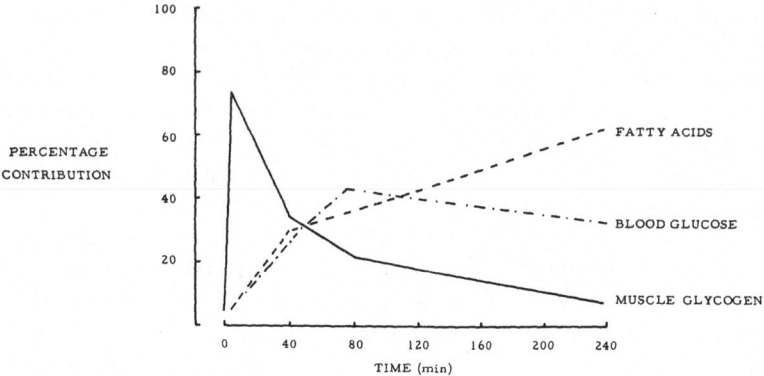

Figure 8.1: Relative Contributions of Various Substrates to Muscle Metabolism during Prolonged Exercise

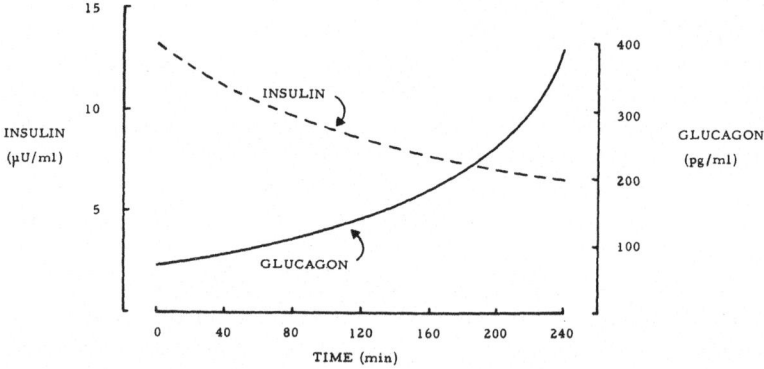

Figure 8.2: Blood Concentration of Insulin and Glucagon During Prolonged Exercise

crine pancreas is able to maintain overall control of nutrient homeostasis albeit with a concentration of blood metabolites set at a higher
value than that which is found in the non-pregnant situation. This situation ensures that the growing feto-placental unit has the nutrients it
requires without an excessive increase in maternal nutrient levels.

In most women the metabolic demands of pregnancy are met by
such adaptive changes in the nutrient homeostatic mechanism and the
needs of both the mother and the growing feto-placental unit are met.
However, in some women the homeostatic mechanism is not able to
adapt sufficiently to the demands and there is a loss of control of blood
metabolite levels. Thus in gestational diabetes, the pancreatic B-cells are
for some unknown reason unable to release sufficient insulin to maintain control of blood metabolite levels and temporary diabetes may
ensue. In this situation the loss of control of metabolite levels may be
sufficient to threaten the growth of the fetus. However, after pregnancy, the diabetes disappears since the B-cells are now able to release
enough insulin to maintain nutrient homeostasis.

Stress

The metabolic response to stress normally involves alterations in the
nutrient homeostatic mechanism which enable blood metabolites to be
elevated above the normal fasting level. These changes appear to be
necessary to ensure that key tissues have adequate metabolite levels to
enable them to make the most appropriate response to the stress. The
metabolic response to periods of acute stress such as fear, anger and
emotional trauma is generally short-lived and involves controlled
increases in blood glucose and free fatty acid levels. The major stimuli
for these changes are adrenaline (released into the circulation from the
adrenal medulla) and noradrenaline (released directly into tissues from
the sympathetic nervous system). These hormones stimulate lipolysis in
adipose tissue and glycogenolysis in liver, effects which lead to increases in the blood level of fatty acids and glucose. In addition, they
exert a direct effect on the pancreatic B-cell and inhibit insulin secretion. This effect is essential if metabolites are to be maintained at an
elevated level since insulin released into the circulation would tend to
lower the level.

The metabolic response to more prolonged periods of stress, such as
occur during shock, depend in part on the severity of the trauma and
may in some instances be associated with a loss of control of blood

nutrient levels. In this situation the persistent presence of catecholamines in the circulation and in tissues, maintains elevated rates of lipolysis and glycogenolysis and may also lead to an increase in the rate of secretion of glucagon from the pancreatic A-cells. In addition, other hormones such as growth hormone may be released into the circulation in response to the trauma. Many of these hormones have an anti-insulin action and since blood insulin levels are already low the situation may arise in which the nutrient homeostatic mechanism fails and there is an uncontrolled mobilisation of stored nutrients. Under these circumstances, in addition to the breakdown of triglyceride and glycogen, muscle protein may also be broken down and the liver may switch to ketone body production with the consequent risk of ketosis. The extent to which these latter changes occur is a reflection of the degree and duration of the shock and it has been suggested that they provide some indication of prognosis for patients in shock, i.e. the more severe the metabolic derangements the worse the prognosis.

Disorders of Pancreatic Endocrine Function

In view of the important role of the islets of Langerhans in regulating nutrient homoeostasis, it is not surprising to find that derangements of pancreatic endocrine function have profound effects on blood nutrient levels. These derangements may involve either the underproduction or overproduction of one or more of the islet hormones and while the overproduction of islet cell hormones is a relatively uncommon clinical problem, the undersecretion of islet hormones, especially insulin, is a major clinical problem.

Pancreatic tumours consisting of relatively pure populations of islet cells have been isolated from a number of patients. The endocrine cells in these tumours have, in general, lost the ability to regulate their rate of hormone synthesis and/or secretion and they tend therefore to secrete relatively large amounts of hormone. Tumours of the B-cell (insulinomas) which release insulin have dramatic effects on the patient even when they are relatively small in size. The episodes of severe hypoglycaemia which are characteristic of this tumour are a reflection of the potency of insulin. Such tumours may be removed surgically when they are large enough to be located or if this should not prove possible, the patient may be treated with streptozotocin, a drug which is relatively selective in its cytotoxicity to B-cells. A number of cases have been described of patients with B-cell tumours which release large amounts of proinsulin (proinsulinomas). In general, the metabolic problems associated with this tumour are less severe than those associated with an insulinoma, a reflection of the relative lack of potency of proinsulin compared with insulin.

The effects of A-cell tumours (glucagonomas) are also less striking than those of insulinomas and not surprisingly the most frequent metabolic finding is that of a raised fasting blood glucose. Since a raised fasting blood glucose is also a characteristic feature of diabetes mellitus the diagnosis of a glucagonoma must be made on the basis of plasma pancreatic glucagon levels.

In the remaining section of this chapter I propose to discuss the acute metabolic problems of diabetes and leave a consideration of the chronic clinical complications of the disease until Chapter 10. In addition, I propose to consider diabetes as being a syndrome composed of two major disease types (type 1 and type 2) each with characteristic metabolic and clinical features. The evidence for this classification will also be discussed in the next chapter.

Type 1 Diabetes

In type 1 diabetes the patient has an absolute lack of functional B-cells in the pancreas. This means that the pancreas cannot produce enough insulin to maintain nutrient homoeostasis and this can only be achieved by the administration of insulin to the patient. In the absence of such exogenous insulin the patient's plasma insulin concentration is usually very low as is the concentration of C-peptide. Thus, in the untreated type 1 diabetic the insulin-requiring processes cannot function normally and those hormones whose actions are normally opposed by insulin can now act unopposed. Not surprisingly therefore there are dramatic changes in the concentrations of many blood nutrients. The time scale of these changes will depend on the rate of loss of B-cell function in the patient who is developing type 1 diabetes, although they occur relatively rapidly (hours-days) when insulin is withheld from a patient with established type 1 diabetes, especially if a normal diet is consumed.

The most characteristic metabolic feature of the untreated diabetic (type 1 or type 2) is fasting hyperglycaemia and this is used as one of the criteria for the diagnosis of the disease. The elevated fasting blood glucose concentration is caused, in part, by the lack of insulin which is required for the uptake and/or utilisation of glucose by tissues such as muscle, liver and adipose tissue. In addition, the unopposed actions of glucagon and possibly other hormones on liver glucose metabolism promote glycogenolysis and gluconeogenesis and inhibit glycogen synthesis, glycolysis and lipogenesis. These alterations in liver metabolism effect-ively switch the liver from a glucose-utilising tissue to a glucose-producing tissue. Thus a combination of increased liver production and decreased peripheral utilisation is responsible for the hyperglycaemia. In the absence of insulin therapy the blood glucose concentration continues to increase and when it exceeds the renal threshold (10mM) glucose appears in the urine (glycosuria). Persistent glycosuria is preceded by post-prandial glycosuria following the ingestion of carbohydrate-rich meals since insulin is unavailable to promote the rapid disposal of the absorbed glucose. The loss of considerable quantities of glucose in the urine necessitates the formation of a large volume of urine (polyuria) and as a consequence the frequency of urination increases as does the need to drink excessive amounts of fluid (polydipsia).

In addition to changes in carbohydrate metabolism there are marked changes in lipid and protein metabolism in the untreated type 1 dia-betic. There is a gradual increase in the rate of breakdown of adipose tissue triglyceride (lipolysis) and muscle protein (proteolysis) as these tissues respond to the lack of insulin and presence of glucagon. This hormonal imbalance also switches liver fatty acid metabolism away from oxidation and/or reesterification to ketogenesis. These changes produce increases in blood fatty acid, amino acid and ketone body levels.

The gradual loss of muscle and adipose tissue mass produces another of the classic features of type 1 diabetes — that of the wasting of body tissues. Certain of the amino acids lost from muscle protein and the glycerol lost from adipose tissue triglyceride provide the substrates for the increased rate of liver gluconeogenesis while the fatty acids released from adipose tissue provide the substrate for the enhanced rate of liver ketogenesis.

In the absence of insulin treatment, ketone body production con-tinues and the blood level of these water soluble metabolites increases (ketosis). When the renal threshold is exceeded ketone bodies appear in the urine (ketonuria) and acetone produced by the chemical breakdown of acetoacetate may be smelt on the breath. The major ketone bodies are strong organic acids and as their concentration continues to increase, the buffering capacity of the plasma begins to fail and the blood pH may drop (acidosis-ketoacidosis).

The changes in blood metabolite levels outlined above occur irres-pective of the intake of nutrients in the diet. Following a meal however, there are even more dramatic increases in blood metabolites, as nutrients such as glucose and amino acids absorbed from the food fail to be utilised by insulin-dependent tissues.

Thus the major metabolic consequences of a lack of insulin in the type 1 diabetic are increases in the blood concentration of glucose, glycerol, fatty acids, amino acids and ketone bodies. The concentration of fatty acids unlike that of glucose and ketone bodies rises to a maximum of 3-4mM. This is because the water-insoluble fatty acids have to be carried in the blood bound to albumin and there is only a limited capacity for this transport. The concentration of glucose and ketone bodies can however increase to very high levels (glucose $\sim$ 70mM, ketone bodies, $\sim$ 30mM) in part becuase of the water solubility of the molecules but also because of the large amounts of protein and triglyceride that can be mobilised to provide the necessary precursors. Unless insulin is administered the patient will die as a result of the effects of the hyperosmolarity and acidity of the plasma on the central nervous system. This catalogue of metabolic disasters can be rapidly curtailed and metabolites returned to normal by the injection of insulin.

Type 2 Diabetes

The metabolic derangements in type 2 diabetes are generally not as severe as those of type 1. This is because the patient has a relatively normal complement of pancreatic B-cells and the fasting plasma insulin levels are normal or may even be elevated. The problem in type 2 diabetes is not an absolute lack of biologically effective insulin, instead it appears to be a relative lack. Thus, under some conditions the amount of insulin released is not sufficient to maintain tight control of blood metabolite levels. This is usually because an as yet undefined defect in the B-cell secretory mechanism prevents the B-cell from responding rapidly enough to a secretory stimulus. Blood metabolites can therefore rise to abnormal levels before there is adequate insulin in the circulation. In addition, in many patients the insulin sensitive tissues have a reduced sensitivity to insulin. This appears to be largely a receptor-defect and may be associated with the obesity which is common in these patients. Indeed, in many patients successful dietary treatment of the obesity may cause the diabetes to disappear.

Since there is always a certain amount of insulin present in the circulation there is not normally the same total loss of control of nutrient homeostasis that is seen with the type 1 diabetic. Indeed ketosis is relatively rare in the type 2 diabetic as is the wasting of muscle and adipose tissue. The major problem appears to occur after a meal when because of the relative lack of biologically effective insulin the blood glucose level rises higher than normal and takes longer to return to the fasting value. This delay may be so long that the fasting value is not reached before the next meal is consumed and there is persistent hyperglycaemia. The clinical consequences of this persistent hyperglycaemia are considered in Chapter 10.

Insulinopathies

Whilst it is known that diabetes is usually caused by an inability to secrete adequate amounts of insulin, it has always seemed likely that in some patients diabetes could be due to the secretion of a structurally abnormal insulin molecule which arose as a result of mutation in the insulin structural gene. A recent case report described a patient who displayed fasting hyperglycaemia (blood glucose of 10mM) in spite of fasting hyperinsulinaemia (100 μU/ml) but who exhibited normal sensitivity to exogenous insulin. Chemical studies revealed that the patient's insulin contained a single amino acid substitution (leucine for phenylalanine) in the active site of the molecule at position B24. This structurally altered insulin had greatly reduced binding to the insulin receptor and as a consequence had a reduced biological potency thus accounting for the diabetic syndrome.

In addition, families have been reported who have increased circulating levels of proinsulin-like material. The defect appears to be in the structural gene for insulin since the proinsulin-like material produced by these patients has an amino acid substitution at the connecting peptide cleavage point which prevents the complete proteolytic conversion of proinsulin into insulin. The mutation can effect either the B-chain/C-peptide (B/C) or the A-chain/C-peptide (A/C) cleavage points and appears to be inherited as an autosomal dominant defect. In the case of the B/C mutation, the secretory product appears to have sufficient biological activity that the patients are able to compensate for their defect by secreting increased amounts of partially cleaved proinsulin and are not diabetic. However, in the case of the A/C mutation this is not so and the patients become diabetic. Thus, while it is possible for diabetes to be caused by a mutation in the insulin structural gene, this is not a major cause of the disease.

Further Reading

Cahill, G.F. Physiology of Insulin in Man. *Diabetes* (1971) *20*, 785-799
Cryer, P.E. Glucose Counterregulation in Man. *Diabetes* (1981) *30*, 262-264
Freinkel, N. Of Pregnancy and Progeny. *Diabetes* (1980) *29*, 1023-1035

Gabbay, K.H. The Insulinopathies. *New England Journal of Medicine* (1980) *302*, 165-167

Given, B.D. *et al.* Diabetes Due to Secretion of an Abnormal Insulin. *New England Journal of Medicine* (1980) *302*, 129-135

Krzentowski, G. *et al.* Glucose Utilization During Exercise in Normal and Diabetic Subjects. *Diabetes* (1981) *30*, 983-989

Newsholme, E.A. The Control of Fuel Utilization by Muscle During Exercise and Starvation. *Diabetes* (1979) *28* suppl. 1, p. 1-7

Pilkis, S.J. *et al.* Hormonal Control of Hepatic Gluconeogenesis. *Vitamins & Hormones* (1978) *36*, 383-460

Robbins, D.C. *et al.* A Human Proinsulin Variant at Arginine 65. *Nature* (1981) *291* 679-681

Unger, R.H. The Milieu Interieur and the Islets of Langerhans. *Diabetologia* (1981) *20*, 1-11

Unger, R.H. *et al.* Insulin, Glucagon and Somatostatin Secretion in the Regulation of Metabolism. *Annual Review of Physiology* (1978) *40*, 307-343

Wahren, J. *et al.* Physical Exercise and Fuel Homeostasis in Diabetes Mellitus. *Diabetologia* (1978) *14*, 213-222

9 DIABETES MELLITUS

Introduction

Primary diabetes mellitus is a disorder of metabolism associated with relative or absolute insulin deficiency. The biochemical and clinical manifestations of the disease span a spectrum from asymptomatic glucose intolerance to symptomatic diabetes with acute metabolic derangements (ketoacidosis, hyperosmolar coma) and with chronic complications (diseases of the vascular, renal and nervous systems). It is an important clinical problem as it currently affects between 1 and 2 per cent of the population of many countries.

Table 9.1: Comparison of Type 1 and Type 2 Diabetes

Feature	Type 1 (IDD)	Type 2 (NIDD)
Proportion of all diabetics (in U.K)	25%	75%
Family history of diabetes	Uncommon	Common
Age at onset	Usually under 30	Usually over 40
Appearance of symptoms	Rapid	Slow
Obesity at onset	Rare	Common
Ketoacidosis	Frequent	Rare
Circulating insulin levels	Low	Low, normal or elevated
Islet B-cells	Markedly decreased	Normal or slightly decreased
Inflammatory cells in islets	Present initially	Absent
Antibodies to islet cells	Present	Absent
HLA association	Yes	No

Two major types of primary diabetes are recognised clinically (Table 9.1). Type 1 or insulin-dependent diabetes (IDD) is characterised by an insulin insufficiency caused by a lack of functional B-cells. Classically, this type of diabetes is first recognised during childhood or adolescence, although the disease can make its clinical appearance at any age. The patient is ketosis-prone and is dependent on exogenous insulin for survival. When untreated, it is associated with the acute symptoms

132

secondary to insulin deficiency (polyuria, polydipsia, polyphagia, weight loss and fatigue).

Type 2 or non-insulin dependent diabetes (NIDD) is characterised by a relative insulin deficiency often related to a defective B-cell secretory response. It is diagnosed most frequently, but not exclusively, in individuals in middle age or older age groups. The patient does not normally require insulin to prevent ketosis, although insulin may be required to prevent hyperglycaemia. This therapeutic goal can, however, often be achieved without insulin, by strict dietary control or by the use of dietary control plus the oral hypoglycaemic drugs.

It is the purpose of this chapter to consider the diagnosis and treatment of diabetes as well as the factors which appear to be involved in the aetiology and pathogenesis of the disorder.

Diagnosis of Diabetes

The diagnosis of diabetes is usually straightforward when there are obvious symptoms such as polyuria, polydipsia, polyphagia, ketonuria and rapid weight loss together with unequivocal elevation of the plasma glucose concentration. In the absence of these signs and symptoms, however, the diagnosis depends on the measurements of plasma glucose concentration under standardised conditions.

The fasting plasma glucose concentration is normally increased in diabetes and a value equal to or above 8mM on more than one occasion is considered diagnostic of diabetes. If the value for the fasting plasma glucose concentration is equivocal then an oral glucose tolerance test may be performed. This measures the ability of the body to clear glucose from the circulation. There are many variables such as diet, emotional state and time of the day that can affect the results of the test and it is necessary to standardise conditions. The test is normally performed on a fasted subject, in the morning, after at least three days of normal diet and physical activity. A venous or capillary blood sample is taken for the determination of the fasting plasma glucose concentration. The subject is then given a drink containing 75 g of glucose and plasma glucose concentrations are determined on blood samples taken at 30 minute intervals in the 2 hour period following the glucose drink. Diabetes is indicated if both the 2 hour sample and one other sample prior to 2 hours have a glucose concentration equal to or exceeding 11.0mM. The oral glucose tolerance test is most frequently used in subjects who are free of the clinical symptoms of diabetes in the hope that early diagnosis and subsequent treatment will slow or prevent the development of the long-term sequelae of the disease.

Some subjects have fasting plasma glucose concentrations above the normal but below the diabetic range, and they often show abnormal, although not diabetic, glucose tolerance tests. These individuals have impaired glucose tolerance and they are at greater risk than the general population of developing diabetes.

Treatment of Diabetes

In all forms of diabetes the amount of insulin released from the B-cell is insufficient to meet the requirement of target tissues for the maintenance of nutrient homeostasis. This imbalance is responsible for the acute metabolic derangements of the disease which may themselves contribute to the chronic clinical complications of the disease. The major objective of all forms of treatment is the normalisation of nutrient homeostasis. The attainment of this goal may prevent or delay the onset of the complications.

Normalisation of nutrient homeostasis requires the supply of and the demand for insulin to be balanced. In type 2 diabetes, obesity and resistance to insulin are common and the demand for insulin is increased in the face of a normal or reduced capacity for B-cell secretion. Treatment is thus directed at reducing the demand for insulin by dietary management. When dietary measures fail or are insufficient, stimulation of endogenous insulin secretion by sulphonylurea drugs may be used. In contrast, type 1 diabetes is characterised by an absolute deficiency of insulin and treatment with insulin, in addition to dietary management, is essential.

Diet Therapy

Dietary management forms the cornerstone of treatment for all diabetic patients. There are two major objectives of diet therapy, the normalisation of blood glucose and the achievement of ideal body weight. These aims are achieved by regulating calorie intake; avoiding carbohydrate-rich food and excesses of fat, and regularity of food intake. Despite the simplicity of these principles, less than 50 per cent of the diabetic population adhere to the recommended dietary regimen.

Regulation of calorie intake is directed at the establishment of ideal body weight. For the type 2 diabetic in whom the incidence of obesity is high, this generally entails a reduction in calorie intake. The importance of weight reduction in these patients is based on the fact that obesity is associated with insulin resistance. This extra demand for insulin disappears when body weight returns to normal. In contrast, the type 1 diabetic, particularly in childhood, generally requires normal amounts of calories to restore body fat and protein and to permit normal growth.

There is no compelling evidence to suggest that the carbohydrate content of the diabetic diet should be disproportionately restricted and there is a tendency to recommend that 50 per cent of the calories should be provided in the form of carbohydrate since there is evidence to suggest that high carbohydrate/low fat diets may increase insulin sensitivity. To avoid marked variations in the blood glucose concentration simple sugars in the form of confectionary and table sugar should be avoided. Instead, the carbohydrate in the diet should be in the form of complex carbohydrates such as starches which are more slowly digested and therefore have a less dramatic effect on the blood glucose concentration.

Day to day regularity of food intake with respect to total calorie and carbohydrate content and with regard to the timing of meals is important in type 1 diabetes since food intake must match the continuing action of injected insulin so as to prevent insulin-induced hypoglycaemia. However, this consistency has to be broken when there is a marked increase in calorie expenditure such as during moderate to severe exercise. Under such conditions extra carbohydrate is taken to meet the needs of contracting muscles and to prevent hypoglycaemia.

Drug Therapy

Two classes of hypoglycaemic drugs are currently employed in the management of type 2 diabetes, the sulphonylureas and the biguanides and they can both be administered orally.

The sulphonylurea drugs share a common molecular structure (Figure 9.1) but have different substitutions on the benzene and urea groups which account for the differences in their potency, metabolism and duration of action. They acutely stimulate insulin release following oral or intravenous administration and this is largely responsible for their hypoglycaemic action. However, when given orally for a period of several months an improvement in glucose tolerance is often observed in the absence of any further increase in plasma insulin levels. This phenomenon is referred to as the 'extra-pancreatic effect of the sulphonylureas' and it may result from the ability of sulphonylureas to increase the number of insulin receptors on the surface of target cells by slowing down their rate of removal. Thus by increasing insulin

(a) Sulphonylureas (general formula)

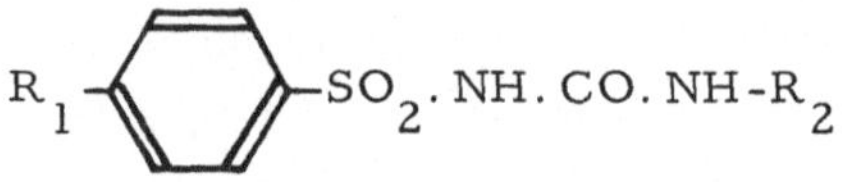

$$R_1 -\!\!\!\bigcirc\!\!\!- SO_2 . NH . CO . NH-R_2$$

(b) Biguanides (phenformin)

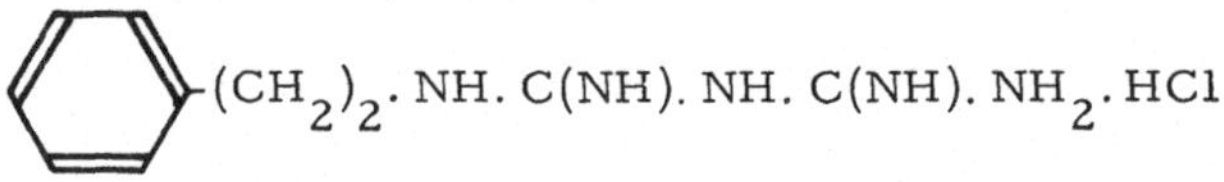

$$\bigcirc\!\!- (CH_2)_2 . NH . C(NH) . NH . C(NH) . NH_2 . HCl$$

Figure 9.1: Structural Formulae of the Oral Hypoglycaemic Agents

binding, sulphonylureas should lead to an increased tissue sensitivity to insulin and to a reduction in the hyperglycaemia.

The biguanides generally have a weaker hypoglycaemic action than the sulphonylureas in man. They are generally not as well tolerated as the sulphonylureas and may produce gastrointestinal disturbances such as anorexia and nausea. In addition, they may enhance the risk of lactic acidosis in diabetics. Unlike the sulphonylureas the biguanides do not stimulate insulin secretion although they do require the presence of insulin for their hypoglycaemic action. This suggests that they may improve the insulin sensitivity of target tissues and the demonstration of an effect of biguanides in increasing the number of insulin receptors on cells would be one mechanism for such an effect.

Insulin Therapy

Only a minority (25 per cent) of the diabetic population require insulin treatment. This minority includes all type 1 diabetics, and type 2 diabetics in whom other therapeutic measures have failed. All patients have their own individual requirement for insulin and treatment must therefore be individualised to meet the overall metabolic, psychological and social needs of the patient. Oral administration of insulin in a form that is resistant to proteolytic destruction is not yet feasible, although liposome-encapsulated insulin may overcome this problem. Insulin has therefore to be administered by subcutaneous or intramuscular injection.

Insulin therapy in an individual patient must be optimised to normalise nutrient homeostasis without causing hypoglycaemia. This is generally assessed by continual monitoring of urine for the appearance of glucose or the measurement of blood glucose concentrations. A number of insulin preparations have been developed each with its own time course of activity. The basic principle involved in the prolongation of insulin activity has been the production of materials with varying solubilities at the pH of the body fluids. The time course of their activity therefore reflects the rate they are absorbed into the circulation from the site of injection. Rapid acting insulins (4-6 hours) consist of relatively soluble microcrystals of zinc-insulin, intermediate acting insulins (up to 12 hours) are mixtures of zinc-insulin and protamine with only limited solubility, and prolonged acting insulins (up to 21 hours) consist of large relatively insoluble crystals of zinc-insulin. These insulins used singly or in combination and in varying amounts can be used to treat all insulin requiring diabetics. However, it is important for the diabetic to be aware of the fact that the rate of absorption of insulin from the site of injection and hence its speed of action depends not only on the type of insulin but also on the site of injection and is increased by exercise. The rate of absorption is fastest from the abdomen, slower from the arm and slowest from the thigh.

The insulin preparations used generally consist of bovine or porcine insulin and many patients develop circulating antibodies to these 'foreign' proteins. However, such antibodies only rarely cause insulin resistance or insulin allergy. Porcine insulin appears to be less antigenic in man than bovine insulin and this possibly reflects the fact that porcine insulin differs from human insulin by only one amino acid residue whereas bovine insulin differs by three amino acids (Table 9.2). Improvements in the fractionation techniques used for the commercial preparation of insulin have significantly increased the purity of all insulin preparations. Although highly purified insulins (more than 99 per cent pure insulin) are only necessary for patients with insulin allergy or lipodystrophy, many clinicians now opt for their use with new patients. Lipodystrophy is a distressing although benign complication of insulin treatment which may take the form of a hypertrophy or atrophy of subcutaneous tissues at the site of insulin injection. Lipohypertrophy is generally observed in patients who repeatedly use the same injection site. The mechanism of the phenomenon is unknown although the fibrous mass slowly regresses when the site of injection is changed. Lipoatrophy commonly occurs in children and adult females within several months of starting insulin therapy and tends to regress after 1-2 years. Lipoatrophy may result from an undetermined contaminant of commercial insulin preparations as the use of highly purified insulin has markedly reduced the incidence of this complication.

Many insulin requiring diabetics are treated each day with a single early morning injection of insulin, although a two injection regimen

Table 9.2: Differences in the Amino Acid Sequences of Human, Bovine and Porcine Insulins

Amino acid location	Human	Bovine	Porcine
A_8	Thr	Ala	Thr
A_{10}	Ileu	Val	Ileu
B_{30}	Thr	Ala	Ala

consisting of an early morning and late afternoon injection of a mixture of rapid and intermediate acting insulins appears to provide a more physiological control of blood glucose concentration. The total amount of insulin that needs to be given each day to a diabetic varies considerably between diabetics but is normally in the range of 20-80 U. This is thought to be similar to the amount of insulin released during 24 hours from the pancreas of a non-diabetic.

Recent development in the fields of chemical and genetic engineering have led to the availability of human insulin for use in the treatment of diabetes. The chemical approach has been to synthesise human insulin from porcine insulin by enzymatically removing the alanine B_{30} of porcine insulin and replacing it with threonine, the amino acid found at B_{30} in human insulin. The amount of human insulin produced by this method will clearly depend on the availability of porcine insulin although the production of human insulin by genetic engineering could theoretically provide us with unlimited amounts of the protein. This would enable the supply of insulin to keep pace with the increasing demand for insulin as the world's population increases and receives better health care. Two approaches have so far been made in trying to apply the developments in recombinant DNA technology to the production of human insulin. The initial approach was to make bacteria synthesise separate insulin A and B chains from chemically synthesised DNA sequences. Following purification, the chains were chemically joined together to form biologically active insulin. The major problem with this approach is the low yield of active insulin obtained when separate A and B chains are joined, since it is difficult to ensure that the correct disulphide bridges are formed. Sufficient insulin has, however, been obtained by this method to enable clinical trials to be undertaken. The results of these trials indicate that human insulin, produced by recombinant DNA technology, has the same biological potency in man as porcine insulin. This result is not altogether surprising since the two proteins differ by only one amino acid. A more promising approach has been to attempt to clone the DNA sequence which codes for the entire proinsulin molecule, since the retention of the C-peptide should ensure the correct folding of the molecule. The recent demonstration by Chan and his co-workers of the synthesis of proinsulin by *Escherichia coli* containing a plasmid with a human preproinsulin cDNA insertion has shown that this approach is feasible. It seems likely, therefore, that recombinant DNA technology will eventually enable human insulin to be produced in large quantities.

Efficacy of Therapy

In healthy normal subjects the blood glucose concentration is one of the most tightly regulated of all metabolic parameters, varying by less than 50 per cent throughout the course of the day. These minimal changes in blood glucose concentration reflect the sensitivity of the feedback regulation between glucose concentration and insulin secretion, as well as the marked sensitivity of target tissues to small fluctuations in insulin. In contrast, in the diabetic, even under optimal conditions of regulated food intake and activity, and multiple daily insulin injections, it is rarely possible to maintain blood glucose concentrations entirely within the normal physiological range for any prolonged period of time. Recognition of this inadequacy has led to the search for newer approaches to treatment. Current research is being directed towards the development of an 'artificial B-cell' and the development of techniques for islet cell transplantation.

Artificial B-Cell

A great deal of effort has been devoted to the development of an artificial B-cell device that is small enough to be implanted into the body of a diabetic. Ideally such a device would constantly monitor the concentration of glucose in the blood or extracellular fluid and would control an automatic insulin delivery unit. The rate of insulin delivery would be determined by the glucose concentration. Successful normalisation of blood glucose concentration has been achieved in insulin-dependent patients by means of such a machine. However, the machines are large and they are attached to the patient via an intravenous connection. The major problems yet to be resolved with this approach are miniaturisation and the development of a glucose sensor which can be implanted directly under the skin.

This type of feedback control system, whilst ideal, is very complicated and much simpler devices have been developed. Normoglycaemia has been achieved in some patients by means of a continuous subcutaneous insulin infusion from a miniature battery-powered syringe pump attached to the patient. These pumps deliver insulin via an indwelling subcutaneous catheter. There are two rates of delivery, the higher rate being designed to be used at meal time. The increased rate of delivery is activated by the patient pushing a small button on the pump and it continues for a fixed period of time, after which the pump automatically returns to the lower rate of delivery. The total dose of insulin given is preprogrammed and in practice is very similar to the amount required by the patient when given as a single daily injection.

Islet Cell Transplantation

Since most, if not all, cases of diabetes reflect a relative or absolute deficiency of insulin, the most rational and efficient therapy would seem to be to replace the deficient endogenous B-cells by new B-cells capable of secreting insulin according to demand. One way of achieving this is to stimulate the patient's own remaining B-cells to replicate until the total cell mass is again sufficient to meet all functional demands. This approach is theoretically possible because B-cell mass has been shown to increase during pregnancy and there must therefore be factors which can stimulate B-cell replication. Identification of the factors which promote B-cell replication may help with this approach.

Alternatively B-cells could be transplanted in sufficient numbers to compensate for the endogenous lack. Numerous investigators have shown that experimental diabetes induced in laboratory animals with chemicals or by removal of the pancreas, can be reversed by injection of islets obtained from healthy, genetically identical animals. Such injections can restore normoglycaemia for as long as 12 months. Islets obtained from genetically non-identical animals are generally rejected within 6 to 12 days unless the recipient is treated with immunosuppressive drugs. These initial studies clearly indicated that islet cell transplantation might provide an answer to the treatment of diabetes. However, there are numerous problems which have to be overcome before islet transplantation becomes feasible in man. These include obtaining sufficient quantities of islets and preventing the rejection of islets obtained from donors.

Exercise

Physical exercise has long been considered beneficial in the treatment of diabetes, since it can reduce the blood glucose concentration and may improve the tolerance to a carbohydrate load. However, in the insulin requiring diabetic the metabolic response to exercise depends on the state of the patient's metabolic control and is, to a large extent, determined by the time interval between insulin administration and the onset of exercise.

In well-controlled patients with mild hyperglycaemia and no keton-aemia the metabolic response is very similar to that of a normal healthy subject. The utilisation of glycogen, blood glucose and free fatty acids by the working muscle is relatively normal (Table 9.3). The glycogen reserves are generally smaller in the diabetic and as a consequence the blood glucose concentration tends to fall as exercise proceeds. However, as in the normal, blood free fatty acid levels increase as exercise proceeds and the exercising muscles switch to fatty acid oxidation thereby reducing their glucose requirement and preventing hypoglycaemia. In addition to the beneficial effect on blood glucose concentration, physical activity improves glucose tolerance and thereby reduces the need for exogenous insulin in the well-controlled diabetic. The improvement in carbohydrate tolerance results from a prolonged stimulation of glucose uptake and glycogen synthesis in tissues following exercise. This may in part be related to an exercise induced increased binding of insulin to muscle cells and to an accelerated absorption of exogenous insulin from sites of subcutaneous injection.

Table 9.3: Percentage Contribution of Various Fuels to the Energy Demand of Working Muscle in Normal and Diabetic Subjects During Exercise Lasting for 2 Hours

Fuel	Normal	Mild non-ketotic diabetic	Ketotic diabetic
Blood glucose	25	28	28
Free fatty acids	27	33	56
Ketone bodies	0	1	6
Muscle glycogen	48	38	10

In contrast, poorly controlled diabetics with marked hyperglycaemia and ketonaemia may respond to exercise with a further rise in blood glucose and ketone body concentration. In addition, there is a more marked increase in the circulating free fatty acid concentration. These responses are largely a reflection of a circulating insulin concentration that is inadequate to prevent excessive lipolysis, gluconeogenesis and ketogenesis, and the extent of the hyperglycaemia and ketonaemia will therefore depend on the interval between insulin injection and the onset of exercise. The rise in free fatty acid concentration follows directly as a consequence of an enhanced rate of lipolysis in adipose tissue. The worsening ketonaemia results from the combination of an elevated circulating free fatty acid concentration and an increased ketogenic capacity of the liver. The increased availability of free fatty acid and ketone bodies in the circulation allows the working muscles to switch rapidly to these fuel molecules and they form the major energy source (Table 9.3) preventing the muscle using glucose in spite of the hyperglycaemia. This occurs at a time when hepatic gluconeogenesis is stimulated and the result is a worsening of the hyperglycaemia.

Aetiology of Diabetes

Since hyperglycaemia and the chronic complications of diabetes are seen in both type 1 and type 2 diabetes, it was for a long time thought that they represented a single disease entity. However, information from a number of studies on the aetiology and pathogenesis of diabetes has accumulated which suggests that it is not a single specific disease entity, but rather a syndrome composed of a number of diseases.

Genetic Studies

It has been known for some considerable time that diabetes tends to run in families. A survey of British diabetics has shown that over 20 per cent had a first-degree relative with diabetes, whereas the rate for non-diabetics was less than 10 per cent. However, since family members usually share the same diet and environment, a familial association of the disease does not necessarily prove that genetic factors are involved.

Studies on the prevalence of diabetes in twins have given some insight into the importance of genetic factors in the development of diabetes. These studies have revealed 45 to 96 per cent concordance (both twins developing diabetes) in monozygotic (identical twins) and only 3 to 37 per cent concordance in dizygotic twins. Since monozygotic twins are derived from the same egg they have identical genetic composition and these results clearly indicate a distinct genetic contribution to the disease. The results of the studies on identical twins have also revealed differences in the genetics of type 1 and type 2 diabetes. Thus when one twin of an identical pair developed type 2 diabetes, the other twin almost always had the disease. However, if one twin developed type 1 diabetes, the other twin developed it within a few years in only half the cases. These findings suggest that genetic factors are predominant in type 2 diabetes but that additional environmental factors are needed to trigger type 1 diabetes.

While it is clear that genetic factors are important in the development of diabetes mellitus, considerable controversy has surrounded explanations of the mode of genetic transmission, and diabetes has been described as 'a geneticist's nightmare'. Genetic analyses by different investigators have implicated practically every mode of inheritance, including autosomal recessive inheritance, autosomal dominant inheritance with incomplete penetrance and multifactorial or polygenic inheritance (i.e. the interaction of multiple gene loci with environmental factors). It is now clear that none of these postulated mechanisms satisfactorily explain the pattern of inheritance in all cases of diabetes. There are a number of reasons for this confusion, chief among which is a lack of knowledge concerning the basic defect(s) underlying the diabetic phenotype. In addition, there is no accurate marker of the diabetic genotype and persons with prediabetes can only be accurately identified retrospectively. Thus, cross-sectional family studies for genetic analysis cannot, as yet, possibly identify all those who will become diabetic in time. Furthermore, it is impossible to control environmental variables such as diet and obesity which may affect the expression of the diabetic genotype.

Finally the major source of error complicating some studies was the failure to recognise genetic heterogeneity in diabetes and data from type 1 and type 2 diabetics were considered together, obscuring possible important differences between these groups. Diabetes is now known to be a genetically heterogeneous group of disorders that share glucose intolerance in common. Heterogeneity implies that different genetic defects or environmental factors (or both) may result in diabetes. Evidence for the genetic heterogeneity of idiopathic diabetes mellitus came initially from the studies on identical twins and more recently from genetic studies of the histocompatibility antigens and from studies on the phenomenon of chlorpropamide-alcohol flushing.

Chlorpropamide-Alcohol Flushing

It has been known for many years that some diabetics treated with chlorpropamide show facial flushing after alcohol. This chlorpropamide-induced alcohol flushing (CPAF) is believed to be genetically determined and is thought to be inherited as a dominant trait. The incidence of CPAF has been reported to be significantly higher in type 2 diabetics than in type 1 diabetics, and it may be possible to use it as a genetic marker for type 2 diabetes.

The mechanism which produces the flush is poorly understood although the demonstration that the flush can be blocked by the specific opiate antagonist naloxone and produced by an enkephalin analogue with opiate-like activity, suggests that enkephalins or other sympathetic neuropeptides might be involved in the response. A rise in met-enkephalin has been shown to occur in response to alcohol and chlor-propamide in normal as well as diabetic subjects and this has led to the suggestion that CPAF is associated with an increased sensitivity to enkephalin. This suggestion is of interest as enkephalin analogues and other neuropeptides have been shown to inhibit insulin secretion. It is possible therefore that the alterations in insulin secretion seen in type 2 diabetes may be related to genetically determined alterations in the sensitivity to neuropeptides. Such a hypothesis must be regarded as speculative at the present time as there is considerable controversy concerning the prevalence, mechanisms and implications of CPAF.

Histocompatibility Antigens in Diabetes

The histocompatibility antigens are the specific marker glycoproteins found on the surface of the nucleated cells of the body which give the cells of each individual their own chemical identity. They are part of an immunological defence system which normally ensures that the healthy cell is not attacked by the body's immune defence system but that it is attacked when it becomes infected with viruses or becomes neoplastic. In addition, they are responsible for the fact that tissue transplanted from one individual to an unrelated individual is normally recognised as 'foreign' and is rejected by the recipient's immune system.

In man the histocompatibility antigens are referred to as the HLA system (human leucocyte antigen). The genes coding for the HLA antigens are on the short arm of chromosome No. 6 and occupy four loci (A, B, C and D) along the chromosome. Each locus has two genes (alleles) one coming from each parent. The alleles at a particular locus may or may not be identical but both are expressed as cell surface proteins. More than 50 different HLA factors have so far been identified and this genetic diversity provides biological uniqueness for every individual in a population. The surface proteins can be identified immunologically and this serves as the basis for the typing of tissues for transplantation. Additional interest in the HLA system was stimulated when it was discovered that certain HLA antigens are found with unusually high frequency in patients with specific diseases including diabetes. Thus, HLA antigens B8 and B15 are two or three times more common in diabetics than they are in non-diabetics and there is an even stronger association between diabetes and the HLA antigens at the D locus (in particular D3 and D4). Moreover, when more than one high-risk allele is present in the same individual the likelihood of developing diabetes is markedly increased. The increased frequency of these HLA antigens is, however, only associated with type 1 diabetes, there being no increase in the frequency of the antigens associated with the type 2 disease. These observations have led to the suggestion that one or more genes in close proximity to the HLA complex along chromosome 6 may be important determinants of type 1 diabetes. The HLA locus on chromosome 6 is intimately involved in the immune response at both the cellular (T-cell function) and humoral/antigen-antibody formation, (B-cell function) levels. It is possible therefore that the high-risk alleles associated with the HLA complex might code for a deficient immune response to agents that preferentially attack B-cells, thereby allowing B-cell damage and diabetes to result. There are, however, other ways in which the immune response could influence the development of diabetes and autoimmunity has been implicated in the pathogenesis of diabetes.

Autoimmunity and Diabetes

Under certain circumstances an individual's immune system can react with its own tissue proteins and cause tissue damage, a phenomenon known as autoimmunity. There is a tendency for patients with one autoimmune disease to develop others, and it has long been known that diabetics have a higher than expected incidence of autoimmune thyroiditis, pernicious anaemia and Addison's disease. It has therefore been suggested that diabetes is one of the family of autoimmune diseases.

The typical pathological findings in type 1 diabetes of lymphocytic infiltration of the islets, together with selective destruction of the B-cells are compatible with an autoimmune lesion. This is also supported by the finding of an antibody that reacts with islet cells in the serum of newly diagnosed type 1 diabetics. The antibody is present in about 85 per cent or more of type 1 diabetics at the time of diagnosis, but decreases to less than 25 per cent after two years. The antibody is rarely present in patients with type 2 diabetes. In addition, lymphocytes from type 1 diabetics often demonstrate significant cytotoxicity against B-cells compared to lymphocytes from non-diabetics, suggesting cell-mediated autoimmunity may be an important component of the pathological process. The demonstration of both cell and humoral mediated immune response to islet tissue in diabetes leaves little room for doubt about the presence of an autoimmune component in the pathogenesis of type 1 diabetes. What exactly triggers the autoimmune reaction is unknown although an environmental factor such as a virus may be responsible.

Viruses and Diabetes

The possibility that virus infection may play a part in the development of diabetes has received considerable attention. In fact, as long ago as 1899 a case was reported in which the onset of diabetes followed an attack of mumps and many clinicians have since commented on the frequency with which the onset of diabetes is preceded by an infection.

The hypothesis that viral infection is one of the causes of diabetes is supported by its abrupt onset, the presence of inflammatory cells in the islets and the destruction of B-cells. In addition, some case reports have shown a temporal relationship between the onset of viral infections such as mumps, rubella ('german measles') and Coxsackie virus B4 and the subsequent development of diabetes. If such infections have an aetiological role, variations in the incidence of diabetes should reflect the epidemiology of the underlying virus infections. There is a pronounced seasonal variation in the onset of diabetes. A low incidence in May and June is followed by a high incidence in autumn and a second more prolonged peak during the winter months. This is very similar to the pattern of childhood illnesses as a whole, many of which are due to viral infections. There is, however, no single known virus or group of viruses with a seasonal incidence resembling that of diabetes. The incidence of diabetes increases with age to a peak at 11-12 years with a smaller peak or shoulder at 5-6 years. Virus infection could provide a basis for such a variation in the age incidence of diabetes since there is a high incidence of viral infections in children at the age of about 5 when they first attend school and at age 11-12 when they transfer to secondary school.

An experimental basis for the viral hypothesis was provided by the demonstration that human islet cells in tissue culture could be infected by Coxsackie B4 and mumps viruses. Replication of the virus in the B-cells was shown to be associated with degranulation of the cells and with subsequent cell death.

Until recently the best evidence that viruses may cause diabetes has come from animal studies. Many diseases of man have counterparts in animals and the study of animal models has contributed to our understanding of human diseases in general. Under certain conditions, variants of several viruses, including the human viruses reovirus 3 and Coxsackie B4 and the animal virus encephalomyocarditis (EMC), are capable of causing diabetes in experimental animals such as the mouse. Diabetes occurs as a result of the loss of B-cell function due to viral attack. The biochemical and pathological features of the disease in mice are very similar to those of diabetes as it occurs in man. The ability of viruses to induce diabetes in mice depends on the strain of the virus and the genetic strain of the animal infected. The B-cell susceptibility to EMC virus attack in mice is under strict genetic control, one or more recessive genes influencing viral uptake and replication. Hereditary factors clearly play an intrinsic part in determining whether or not B-cells are injured by a particular virus.

The first direct evidence that viruses are capable of causing diabetes in man was published in 1979 when it was shown that the dramatic onset of diabetic ketoacidosis in a 10-year-old boy was secondary to B-cell damage caused by a viral infection (Yoon *et al.*). The virus was isolated from the boy's pancreas and shown to be a diabetogenic variant derived from Coxsackie B4. Diabetes is not a common consequence of Coxsackie B4 infection since about half the population are at some time infected with this virus but less than 1 per cent develop diabetes. The B-cells of this patient may, however, have been particularly susceptible to viral attack because of genetic factors. Thus it appears likely that viruses are one of the environmental factors that play a role in the development of diabetes although it is clear that their effect only becomes manifest in certain genetically predisposed individuals, or in individuals with a depleted B-cell reserve caused by a series of viral infections or other environmental insults such as chemicals, drugs or toxins which induce B-cell damage.

Drugs and Toxic Chemicals and Diabetes

The drugs alloxan and streptozotocin are known to be toxic to the B-cell and are capable of causing B-cell destruction. The effects of these drugs are relatively specific for the B-cell and both agents are widely used to induce experimental diabetes in animals. Paradoxically, because of its specificity and toxicity, streptozotocin has also been used successfully in man to treat B-cell tumours (insulinomas). The rodenticide, Vacor, has also been shown to cause B-cell destruction and diabetes when ingested by man. There is a strong possibility, therefore, that there may be other chemicals in the environment which can exert a selective toxic effect on the B-cell and thereby induce diabetes.

Role of Glucagon in Diabetes

It has been suggested that the metabolic derangements of diabetes are a consequence not only of abnormal B-cell function but also of abnormal A-cell function. This controversial proposal has led to numerous studies being performed with the aim of defining the role of the A-cell in diabetes. Studies *in vitro* have shown that glucagon has the property of increasing the mobilisaton of glucose, free fatty acids and ketone bodies, metabolites which are found in excessive concentrations in the blood of diabetic patients, while studies *in vivo* have shown that exogenous glucagon administration to insulin-deficient diabetic subjects causes a significant rise in plasma glucose, free fatty acids and ketone bodies, aggravating the metabolic abnormalities of diabetes. Thus, glucagon possesses properties which qualify it as a potential diabetic hormone. In addition, there is general agreement that plasma glucagon levels are elevated in diabetes mellitus, the highest values being recorded in the extreme situation represented by keto acidosis.

In order to determine directly the role of glucagon in diabetes, the effect of the removal of glucagon from the circulation on the development of the metabolic abnormalities of diabetes has been investigated. Removal of glucagon from the circulation has been achieved using totally pancreatectomised subjects or by inhibiting endogenous glucagon secretion in diabetic subjects by somatostatin infusion. These studies have shown that, while the metabolic abnormalities of diabetes can develop to a limited extent in the absence of circulating glucagon, nevertheless elevated glucagon levels do play an important role in the development of the marked hyperglycaemia and ketosis seen in uncontrolled diabetes.

The elevated plasma glucagon levels seen in diabetes arise as a consequence of an abnormal secretory response of the A-cell which becomes relatively insensitive to glucose. The ability of glucose to inhibit glucagon secretion normally depends on the presence of insulin since the A-cell appears to require insulin for glucose entry and the subsequent inhibition of glucagon release. The question has therefore been asked as to whether the abnormalities of glucagon secretion seen in diabetes may be a consequence of relative or absolute insulin deficiency. In type 1 diabetes insulin therapy usually reduces basal glucagon levels and in these patients it is possible that glucagon excess may result in part from insulin lack. However, in type 2 diabetes, abnormalities in glucagon secretion exist, despite the persistence of insulin. The mechanisms leading to these abnormalities are unknown, although an intrinsic derangement of the glucagon releasing mechanism, a defective release of somatostatin or reduced sensitivity of the A-cell to insulin are all possibilities.

Thus, it appears likely that abnormal B-cell function plays the major role in the development of the metabolic abnormalities seen in diabetes, although abnormal A-cell function may contribute to the acute metabolic abnormalities seen in uncontrolled diabetes.

Somatostatin in Diabetes

Hyperplasia of D-cells and increased islet content of somatostatin have been reported in type 1 diabetes and these observations have led to the suggestion of a possible involvement of the D-cell in the pathological process. However, it is not known whether these changes contribute to, or result from the pathological process. In addition, it is not known whether they indicate excessive or diminished secretion of somatostatin. Conceivably a primary deficiency of somatostatin could result in excessive secretion of insulin and glucagon and, ultimately to hyperplasia of both A and B-cells. In susceptible individuals lack of appropriate B-cell hyperplasia or a defective insulin secretory mechanism could lead to a syndrome resembling type 2 diabetes. Conversely excessive secretion of somatostatin could cause diabetes by inhibiting the secretion of insulin. Diabetes associated with hypoinsulinaemia has in fact been observed in patients with somatostatin-producing tumours of the pancreatic D-cells. At the present time, however, there is insufficient evidence to suggest an important role for D-cells in the development of diabetes.

Pathogenesis of Diabetes

The results of studies on the aetiology of diabetes have shown that diabetes arises as a result of a complex interaction between the individual and the environment and that some individuals may be genetically predisposed towards the disease. In type 1 diabetes the result of this interaction is the relatively specific destruction of pancreatic B-cells, while in type 2 diabetes the result is an impairment of the pancreatic B-cell's ability to secrete insulin rapidly in response to hyperglycaemia. It is likely that the basis of the genetic predisposition and the nature of the environmental factors involved differ in the different types of diabetes.

In type 1 diabetes the environmental factors so far suggested include viruses and toxic chemicals. These agents appear to interact specifically with the B-cells of certain individuals. As a result of this interaction the B-cell is either irreparably damaged or is altered sufficiently for the body's immune defence system to attack and destroy the cell. In addition, since the lost B-cells are not subsequently replaced it appears that the B-cells of these individuals have an impaired ability to replicate. If people who were genetically predisposed to type 1 diabetes could be identified, it might be possible to protect them from environmental hazards such as viruses or to reduce the intensity of their autoimmune response to such factors. These possibilities may become reality in the future.

In type 2 diabetes the environmental factors may be dietary and involve an intake of calories in excess of the body's requirements since it is often associated with obesity. The intake of excess calories together with the peripheral resistance to insulin that occurs in obesity requires an adaptive increase in the secretory capacity of the pancreatic B-cell. However, in some individuals such an adaptive response is not possible or is short-lived. Insulin levels are then inadequate to maintain normoglycaemia and diabetes results. In other individuals, it appears that the adaptive change of the B-cell secretory mechanism eventually results in permanent damage to the secretory mechanism such that it is unable to respond rapidly to hyperglycaemia. The damage appears to occur in the regulatory systems controlling the insulin secretory rate rather than the secretory mechanism itself since the sulphonylureas can restore the B-cell secretory function to normal. There is a great deal of evidence to suggest that alterations in the cyclic AMP system of the B-cell is responsible for the long-term adaptive changes in the secretory response of this cell to a variety of physiological situations. It may be therefore that it is a defect in the cyclic AMP system of the B-cell that is responsible for the impaired secretory response seen in type 2 diabetes. Evidence in favour of this hypothesis has come from the observation that the sulphonylureas increase insulin secretion, at least in part by an effect on cyclic AMP levels in the B-cell.

In addition, studies of the effects of pregnancy on the B-cell also support this possibility. Part of the metabolic adaptation to pregnancy involves an increased responsiveness of the B-cell to stimulation by glucose and this normally produces an elevation in the circulating insulin concentration. Elevated circulating insulin levels are necessary during pregnancy as maternal tissues such as adipose tissue become less sensitive to insulin, due in part to the high circulating levels of placental lactogen. The response of the B-cell to pregnancy involves an increased activity of adenylate cyclase and a raised intracellular cyclic AMP level which enables the B-cell to secrete insulin at an increased rate. However, the B-cells of some women fail to develop an increased responsiveness and they are unable to meet the extra demand for insulin. Their plasma insulin levels become inadequate to maintain blood glucose levels and they become diabetic. After pregnancy when the demand for extra insulin has gone the diabetes disappears. This type of diabetes which only occurs during pregnancy is termed gestational diabetes. It is usually mild, although in some cases it can be severe enough to require insulin therapy. It is possible that the failure of the B-cell to respond in gestational diabetes may reflect an impairment in the adenylate cyclase activity of this cell type since this enzyme appears to be important in mediating the normal B-cell response to pregnancy.

The results of the CPAF studies which suggested that the impaired B-cell secretory activity seen in type 2 diabetes might result from an increased tissue sensitivity to neuropeptides are also compatible with this theory. Thus it is likely that these peptides inhibit insulin secretion via an inhibition of the adenylate cyclase activity of the B-cell and in the presence of reduced cyclase activity they would become relatively more effective at lowering cyclic AMP levels and inhibiting insulin secretion.

It is possible therefore that the basic defect in the B-cell of type 2 diabetics is an impairment of the cell's adenylate cyclase activity. Identification of this and possibly other defects in the B-cell may lead to the development of new drugs which can overcome the defect. Until that time comes many type 2 diabetics will have to rely on the sulphonylureas to improve the secretory response of their damaged B-cells.

Further Reading

Cahill, G.F. & McDevitt, H.O. Insulin-dependent Diabetes Mellitus: The Initial Lesion. *New England Journal of Medicine* (1981) *304*, 1454-1465

Chan, S.J. *et al*. Biosynthesis and periplasmic segregation of human proinsulin in *Escherichia coli*. *Proceedings of the National Academy of Sciences (USA)* (1981) *78*, 5401-5405

Cudworth, A.G. Type 2 (Insulin-Independent) Diabetes — Fibres and Flushers. *Diabetologia* (1979) *17*, 67-69

Lebovitz, H.E. & Feinglos, M.N. Therapy of Insulin-Independent Diabetes Mellitus. *Metabolism* (1980) *29*, 474-482

Lernmark, A. & Baekkeskov, S. Islet Cell Antibodies. *Diabetologia* (1981) *21*, 431-435

Mann, J.I. Diet and Diabetes. *Diabetologia* (1980) *18*, 89-95

Miller, W.L. & Baxter, J.D. Recombinant DNA — A New Source of Insulin. *Diabetologia* (1980) *18*, 431-436

National Diabetes Data Group. Classification and Diagnosis of Diabetes Mellitus and Other Categories of Glucose Intolerance. *Diabetes* (1979) *28*, 1039-1057

Pyke, D.A. Diabetes: The Genetic Connections. *Diabetologia* (1979) *17*, 333-343

Rayfield, E.J. & Seto, Y. Viruses and the Pathogenesis of Diabetes Mellitus. *Diabetes* (1978) *27*, 1126-1136

Rizza, R.A. *et al.* Control of Blood Sugar in Insulin-Dependent Diabetes. *New England Journal of Medicine* (1980) *303*, 1313-1318

Sutherland, D.E.R. Pancreas and Islet Transplantation I & II. *Diabetologia* (1981) *20*, 161-185 & 435-450

Vranic, M. & Berger, M. Exercise and Diabetes Mellitus. *Diabetes* (1979) *28*, 147-160

Watkins, P.J. Congenital Malformations and Blood Glucose Control in Diabetic Pregnancy. *British Medical Journal* (1982) *284*, 1357

Wilkin, J.K. Chlorpropamide-Alcohol Flushing, Malar Thermal Circulation Index, and Baseline Malar Temperature. *Metabolism* (1982) *31*, 948-958

Yoon, J.W. *et al.* Virus-Induced Diabetes Mellitus. *New England Journal of Medicine* (1979)*300*, 1173-1179

10 DIABETIC COMPLICATIONS

Introduction

At the beginning of this century a young diabetic was lucky to survive two years from the time of diagnosis. The advent of insulin dramatically changed this gloomy prognosis, but as diabetics survived longer, other devastating complications of their disease became apparent. The complications of diabetes are a heterogeneous group of clinical disorders which can affect the vascular system, the kidney (nephropathy), the eye (retinopathy), the nervous system (neuropathy) and possibly other tissues. At present the development of any of the disorders in a given diabetic patient is unpredictable and not all diabetics, even those of very long duration, develop the complications. The frequency of specific complications may vary in type 1 and type 2 diabetes, but none of the common complications are restricted to either type.

Considerable controversy exists concerning the questions of whether the complications are the result of the metabolic disturbances caused by insulin deficiency or whether they, like the metabolic disturbances, are symptoms of an underlying generalised tissue defect. This question is of enormous importance to those who treat diabetics. If the long-term, life-threatening complications of diabetes could be shown to arise as the result of the metabolic disturbances, then the objective of all diabetic therapy must be to maintain blood metabolite concentrations within the physiological range at all times. If however, it could be shown that strict control of blood metabolite concentrations does not prevent or reduce the frequency of complications then there would seem to be little point in elaborate therapeutic measures aimed at strict control.

It is the purpose of this chapter to consider what is known concerning the molecular basis of the complications of diabetes. In addition, an attempt will be made to present the evidence in favour of the suggestion that strict control, i.e. control within the physiological ranges of blood glucose concentrations, has a significant effect on the amelioration of diabetic complications.

Diabetic Neuropathy

Diabetes may affect the nervous system producing damage to both the peripheral and autonomic nerves. The most frequent neurological problem is peripheral neuropathy which typically involves loss of sensation particularly of the lower extremities. Autonomic neuropathy on the other hand may produce symptoms such as postural hypotension and impotence. All these symptoms are the clinical expression of structural and biochemical alterations in the axon and the myelin sheath which cause alterations in nerve electrophysiology. Changes in nerve electrophysiology such as decreased sensory and motor nerve conduction velocities can be detected in many newly diagnosed diabetics and nerve function often improves after hyperglycaemia is corrected by treatment. These observations suggest that the changes in nerve function which occur in diabetes may be secondary to the metabolic changes.

Recent evidence suggests that the axon is the initial site of damage in the peripheral nerve, with shrinkage of the axons and expansion of the endoneural space preceding subsequent Schwann cell disease and segmental demyelination. The molecular basis of these alterations is not clear but it may involve the accumulation of sorbitol, which has been observed in the nerves of diabetics and which may have osmotic or toxic effects. In addition, there are other metabolic changes in the nerve, such as changes in myelin lipid composition, reduction in tissue myo-inositol and increased glycosylation of neural proteins.

The observation that nervous tissue from diabetics contains reduced concentrations of myo-inositol and its phospholipid derivatives has led to the suggestion that the functional disorders of diabetic nerves may be related to abnormalities in the axonal metabolism of this substance.

Myo-inositol is a 6-carbon cyclic polyalcohol which is normally obtained largely from the diet. It is present in peripheral nerves at a concentration of $\sim 3\text{mM}$ (approximately 30 times that of its concentration in plasma) and forms an important component of membrane phospholipids involved in nerve impulse transmission. The breakdown of myo-inositol containing phospholipids increases dramatically during nervous conduction and synaptic transmission and their subsequent resynthesis requires the presence of myo-inositol (Figure 10.1). However, the low concentration of myo-inositol observed in diabetic nerves may impair the resynthesis of myo-inositol containing phospholipids since the enzyme responsible for the addition of myo-inositol has a high K_m for inositol. These observations may account for the impaired nerve function in diabetics. The reason for the low myo-inositol content of

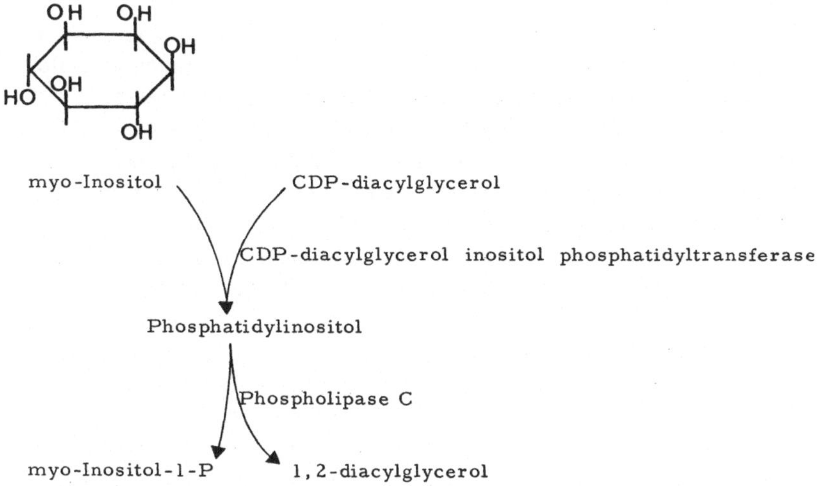

Figure 10.1: Myo-inositol Metabolism

diabetic nerves is not known although it may be one of the consequences of an inadequate supply of nutrients to the nerve since there are marked changes in the structure and function of the vascular system in diabetics which could affect the nutrient arterioles supplying the nerves.

Diabetic Vascular Disease

Vascular disease is one of the commonest, as well as one of the most serious chronic complications of diabetes and diabetics are susceptible to disease of both the large muscular arteries (macroangiopathy) and the capillaries (microangiopathy).

Macroangiopathy

The major large vessel problem in diabetics is atherosclerosis, and this can affect the myocardium, brain and lower limbs. The incidence of atherosclerosis in diabetics is significantly higher than in non-diabetics, although it has a similar distribution and appearance in both groups. While the presence of hyperlipidaemia, hypertension, obesity or smoking will contribute to the development of atherosclerosis in diabetics and non-diabetics, epidemiological studies indicate that there are additional factors associated with the disease in diabetics. In addition, the fact that many diabetics can survive their disease for long periods without developing vascular disease suggests that environmental and/or genetic factors are also important in the aetiology of the disease.

Abnormalities in lipid and lipoprotein metabolism are associated with atherosclerosis in non-diabetics, and since abnormalities of lipoprotein metabolism have been reported in both type 1 and type 2 diabetes they are likely to be a major contributory factor to the development of diabetic macroangiopathy. Thus elevated levels of VLDL triglyceride and LDL cholesterol have been found in many diabetics while HDL cholesterol, which varies inversely with cardiovascular disease risk in non-diabetics, appears to be decreased in some diabetics. It is not known to what extent these changes in lipid metabolism in diabetes are reflections of the metabolic abnormalities.

Studies on the development of atherosclerosis in the arterial wall have led to a conceptual model for the pathogenesis of macroangiopathy. According to this model, the initial event is injury to the arterial endothelium which allows blood components such as platelets and lipoproteins to enter the subendothelial space. These components may be partly responsible for the subsequent subendothelial proliferation of smooth muscle cells and the deposition of lipid and collagen that ultimately produces the atheromatous plaque. Many of the elements of this proposed pathogenic sequence appear to be accentuated in diabetes.

Thus the increased permeability of the endothelium observed in diabetes may permit the passage of blood components into the subendothelial space, while the subsequent interaction of platelets with subendothelial structures may be enhanced by the increased adhesiveness and aggregation of platelets from diabetics. In addition, there may be factors in diabetic serum which promote smooth muscle proliferation and stimulate arterial wall collagen production and the accumulation of excess lipid may be related to changes in low density lipoprotein metabolism that have been reported in diabetes.

It has been suggested that insulin may play a role in the development of atherosclerosis in diabetes since it promotes the accumulation of lipids in the arterial wall and it promotes the proliferation of arterial wall smooth muscle cells. In this context it is important to realise that few diabetics are truly insulin deficient. The majority of type 2 diabetics have relatively high basal insulin levels while type 1 diabetics receive doses of insulin in excess of the normal daily insulin output of the pancreas. It is possible therefore that the increased incidence of atherosclerosis in diabetics is related in part to abnormal circulating concentrations of insulin.

Microangiopathy

Diabetic microangiopathy involves alterations in the microvasculature that profoundly alter the normal function of the kidney (nephropathy), retina (retinopathy) and other tissues such as the heart. These alterations may include disseminated intravascular coagulation and diffuse platelet microthrombi, both of which impair blood flow. In addition, there is an increase in the thickness of the capillary basement membrane, an alteration which may be responsible for the changes in the permeability of the membrane that have been observed in diabetes. Most of the evidence currently available suggests that microangiopathy is a true consequence of the primary metabolic disturbances of diabetes. Thus, thickening of the basement membrane follows the onset of the metabolic disturbances and progresses with the duration of the disease. In addition, it is less common in well-controlled diabetics and it also occurs in secondary diabetes. The normal structure of the vascular system depends in part on complex interactions between the blood vessel wall, erythrocytes, platelets and plasma factors. Alterations in several of these components have been observed in diabetics and this has led to speculation concerning the role of these alterations in the development of the vascular complications of the disease.

Intravascular coagulation is sometimes a feature of diabetic microangiopathy and this may be related to an increase in the circulating level of clotting factors in the diabetic. In particular, fibrinogen levels and factors V, VI, VIII and von Willebrand factor have all been shown to be increased in diabetes, while there appears to be a lower tissue and systemic fibrinolytic activity in diabetes, a feature which would clearly prevent the rapid removal of blood clots. Diffuse platelet microthrombi are also found in diabetics and abnormalities of platelet function have been described in some diabetics. Thus, platelets from diabetics may show increased adhesiveness, increased aggregation and excessive release of thrombogenic platelet factors. Furthermore, platelet disaggregation appears to be reduced in diabetes. Poorly controlled diabetes is also associated with increased plasma viscosity related to increased erythrocyte aggregation. The resistance to flow offered by such viscous blood could lead to increased hydrostatic pressure in arterioles, altering the permeability of the basement membrane and resulting in the extravasation of plasma proteins. Such changes may be important in the eye and kidney. Finally, there are a number of erythrocyte abnormalities in diabetes which may lead to decreased tissue oxygenation and it has been suggested that relative tissue hypoxia may play a role in the development of several diabetic complications. Thus, the increased aggregation of erythrocytes and their reduced deformability that have been observed in diabetes may impair their passage through the smaller capillaries. In addition, the increased levels of glycosylated haemoglobin and decreased effective levels of 2,3-diphosphoglycerate which have also been observed in such erythrocytes would tend to increase the affinity of haemoglobin for oxygen and decrease the delivery of oxygen to critical tissues.

Diabetic Nephropathy

Diabetic microangiopathy often affects the kidney resulting in a progressive loss of renal function and this is characterised clinically by proteinuria and a decreasing glomerular filtration rate. In some cases severity of these clinical features appears to correlate with the thickness of the basement membrane of the glomerular capillaries. Thus, at the onset of diabetes the membrane has a normal thickness but as the disease progresses and renal function decreases there is a corresponding increase in membrane thickening. Ultimately the continual thickening of the capillary basement membrane combined with the expansion of the mesangium leads to the occlusion of the glomerulus and results in chronic renal failure. The progression and extent of these changes appear to be a function of the duration and degree of the metabolic derangements suggesting that they may develop as a consequence of the metabolic derangements.

The basement membrane is composed of collagen-like glycoprotein material that is rich in the amino acids glycine, hydroxyproline and hydroxylysine. Studies have indicated that there is an increased amount of basement membrane protein in glomeruli from diabetics and that this material has an altered composition with an increased presence of glycosylated hydroxylysine-rich subunits. This compositional change might be partly responsible for the altered filtration properties of the glomerular basement membrane seen in diabetes since increased glycosylation may interfere with polypeptide packing and/or hydroxylysine cross-link formation. Such alterations might increase the effective pore size of the basement membrane allowing the loss of material which would normally be retained. However, changes in permeability could also arise as a result of increased hydrostatic pressure in the glomerulus caused by the resistance to flow offered by viscous blood in the efferent arteriole.

The progressive fall in glomerular filtration rate observed in many diabetics leads eventually to chronic renal failure and appears to be largely a consequence of the gradual occlusion of the glomerular capillaries caused by the accumulation of excessive amounts of basement membrane material. This accumulation might represent increased synthesis, decreased degradation or a combination of both. Evidence has accumulated which suggests that there is an increased rate of glomerular basement membrane polypeptide synthesis in diabetes and that hyperglycaemia may play an important role in stimulating this process. In addition, it is possible that increased glycosylation of the membrane proteins may reduce their susceptibility to proteolysis and thereby reduce their rate of degradation.

Diabetic Retinopathy

Impaired vision in patients with diabetes can result from cataracts, refractive errors and glaucoma. However, more serious and much more common are the microvascular changes (retinopathy) that may occur in the eyes of diabetics. It is possible to distinguish two stages in the development of severe retinopathy. Initially, alterations occur in the retinal vessels (non-proliferative or background retinopathy), which result in the breakdown of the blood-retinal barrier with consequent leakage of vessel contents into the retina. This may be followed by proliferative retinopathy where abnormal new vessels may haemorrhage and form fibrous tissues resulting in greater visual impairment and even blindness.

It has been suggested that regional ischaemia may be an important factor in the development and progression of diabetic retinopathy. Thus, initially there is a breakdown of the blood-retinal barrier with a resultant accumulation of fluid and vascular components in the retina. These changes are thought to significantly reduce oxygen diffusion in the retina producing areas of ischaemia. In addition, the haematological abnormalities of diabetes discussed previously, would tend to result in a reduced oxygen delivery to retinal tissue. The response to this ischaemia is thought to involve a compensatory vasodilation of the retinal vessels which increases blood flow through the retina. This response, however, gradually becomes limited because of increasing focal occlusion of the capillaries caused in part by progressive basement membrane thickening and endothelial cell growth. This progressive capillary occlusion may lead to a worsening of the regional ischaemia and may promote the growth of new vessels.

In early diabetes, the breakdown of the blood-retinal barrier appears to be related to a loss of tight junctions between the endothelial cells. The degree of breakdown correlates with the extent and duration of the loss of metabolic control and normalisation of blood glucose levels may significantly reduce the increased capillary permeability.

Diabetic Cataracts

Cataracts (opacity of the ocular lens) are found more frequently in diabetics than non-diabetics and in some type 1 diabetics their development may be particularly rapid. The opacity of the lens is thought to result from the formation of high molecular weight (50-200 $\times$ 10^6 dalton) aggregates of lens proteins which scatter light rather than allowing the light to penetrate the lens. The aggregation of lens proteins appears to involve the formation of disulphide bonds between protein molecules and is facilitated by an increased susceptibility of protein sulphydryl groups to oxidation. The increased glycosylation of lens proteins that occurs in diabetes appears to increase their susceptibility to oxidation although the major factor responsible for maintaining sulphydryl groups on proteins and preventing their oxidation to disulphide bonds, is the level of reduced glutathione in the cell. This in turn is regulated by the availability of NADPH (Figure 10.2). It is likely that in the presence of severe hyperglycaemia the level of NADPH in the diabetic lens will be low since under these conditions the lens converts excess glucose into sorbitol at the expense of NADPH (Figure 10.3). This low level of NADPH during hyperglycaemia may predispose the sulphydryl groups of lens proteins to disulphide bond formation with the resultant formation of cataracts.

Diabetic Pregnancy

Pregnant diabetic women are at greater risk of fetal loss with or without congenital malformations than non-diabetic women and babies born at term to diabetic mothers are often larger than those born to non-diabetic mothers. However, the risk of these problems has been shown to be considerably reduced by good obstetric management and by strict control of maternal blood glucose levels. Indeed, it has been shown that maternal hyperglycaemia exerts profound effects on the fetus and may be at least partly responsible for the complications.

Organogenesis occurs during the first eight weeks of pregnancy and hyperglycaemia during this period appears to be related to the develop-

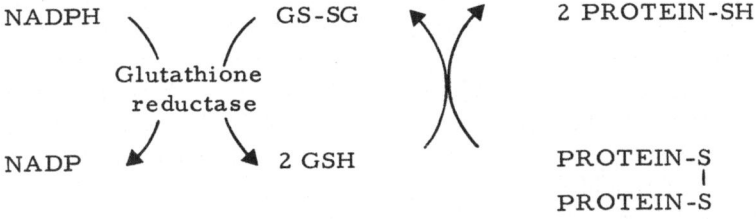

Figure 10.2: The Role of Glutathione in the Maintenance of Protein Sulphydryl Groups. GSH = reduced glutathione; GS-SG = oxidised glutathione.

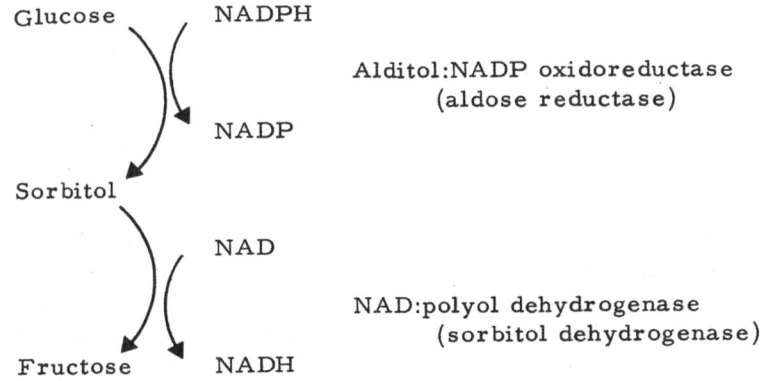

Figure 10.3: Sorbitol Metabolism

ment of congenital malformations. In addition, maternal hypergly-caemia causes fetal hyperinsulinaemia and this is, at least partly, respon-sible for increased deposition of fat in the fetus. Fetal hyperinsulinism may also be responsible for the delayed lung maturation with resulting respiratory distress seen in some infants. It also causes neonatal hypo-glycaemia, a condition common in the newborn of diabetic mothers. Thus, many of the problems of pregnancy in diabetics appear to be related to the metabolic abnormalities of the mother.

Causes of the Complications of Diabetes

The weight of evidence currently available suggests that many of the clinical complications of diabetes are the direct result of the metabolic disturbances caused by relative or absolute insulin deficiency, although it is possible that independently determined genetic and environmental factors can alter the likelihood of developing a specific complication. It is to be hoped therefore that therapeutic measures will be developed that will enable the concentration of glucose and other metabolites in the blood and tissues to be controlled within the physiological range. In this way it should be possible to reduce the incidence and severity of the complications. In this context it is important to determine how the metabolic disturbances lead to the development of the complications. Progress in this area has been slow owing to the fact that the normal biochemistry of the affected tissue or cell type is poorly understood. However, certain common features are beginning to emerge. The most striking and consistent metabolic abnormality of diabetes is hyperglycaemia and several recent studies have suggested ways in which hyperglycaemia may lead to the development of many of the complications.

Many of the complications of diabetes involve tissues, such as peripheral nerves, blood vessels and lens tissue, that have a glucose uptake mechanism which does not require insulin. In these tissues the intracellular glucose concentration is determined largely by the extracellular concentration which is increased in diabetes. A high intracellular glucose concentration is likely to be associated with alterations in the normal metabolism of glucose. Several such alterations have been demonstrated in tissues from diabetic patients. In particular, glycosylation-reactions have been shown to occur more readily in diabetics and the metabolism of glucose via alternative pathways such as the polyol pathway is increased. These aspects of glucose metabolism will be considered further as they may provide a clue to the metabolic basis of diabetic complications.

Polyol Pathway

In diabetes, specific intermediates of glucose metabolism such as sorbitol and fructose accumulate to high concentrations in nervous tissue, capillary walls and the lens of the eye. These intermediates are formed by the activity of the polyol (sugar alcohol) pathway which operates irreversibly in intact tissues (Figure 10.3). The pathway catalyses the reduction of glucose to its polyol derivative, sorbitol, and its subsequent oxidation to D-fructose, and involves the sequential action of alditol: NADP oxidoreductase and NAD:polyol dehydrogenase. High ambient glucose concentrations stimulate the activity of the pathway and lead to the accumulation of sorbitol and fructose and a reduction in tissue

NADPH levels. The accumulation of sorbitol and fructose may lead to changes in osmotic pressure causing the cells to swell and thereby damaging the tissue. In addition, the intermediates may themselves be toxic when present in high concentrations, thereby causing cell and tissue damage. The reduction in intracellular NADPH levels may also lead to cell damage as the integrity of the plasma membrane depends on the presence of reduced sulphydryl groups on membrane proteins and NADPH is necessary to maintain these groups in the reduced form (Figure 10.2). The polyol pathway may thus provide a link between hyperglycaemia and tissue damage in diabetes, although its exact role in the pathogenesis of the complications of diabetes has not yet been established.

Glycosylation

Recent studies have shown that under certain conditions glucose will combine non-enzymatically with free amino groups in proteins to produce glycosylated proteins. The reaction appears to proceed through the formation of a Schiff base between the aldehyde form of the sugar and free amino groups in protein. This is followed by Amadori rearrangement of the Schiff base to a ketoamine derivative which is stabilised by cyclisation and formation of the hemiketal structure (Figure 10.4).

Interest in this reaction was stimulated by the observation that erythrocytes from diabetic patients contained a higher proportion of a glycosylated haemoglobin (A_{1c}) than normal. Haemoglobin A_{1c} is one of three negatively charged minor haemoglobin components that can be separated from haemoglobin A by ion-exchange chromatography. In normal individuals haemoglobin A_{1c} is present as 3-5 per cent of the total haemoglobin and in diabetics this can increase to as much as 20 per cent. Haemoglobin A_{1c} differs from haemoglobin only in the attachment of 1-amino-1-deoxyfructose to the amino-terminal valine of each β-chain. Glycosylation of haemoglobin reduces the reactivity of the molecule with 2,3-diphosphoglycerate which increases the affinity of the molecule for oxygen. The irreversible glycosylation of haemoglobin A to form haemoglobin A_{1c} occurs slowly throughout the 120 day lifespan of the red cell at a rate proportional to the blood glucose concentration. The steady-state value of haemoglobin A_{1c} in the blood reflects a balance between destruction of old cells with haemoglobin A_{1c} and the production of cells without haemoglobin A_{1c}. It is also a function of the time-averaged blood glucose concentration during the preceding few weeks. Measurement of haemoglobin

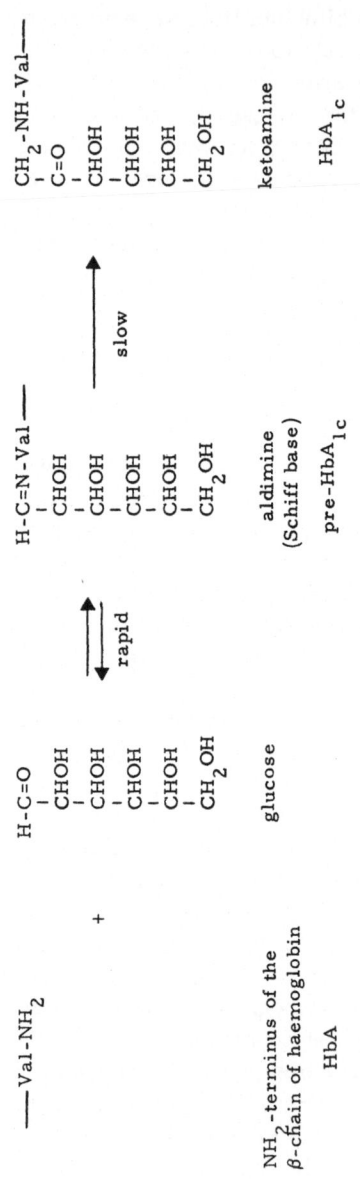

Figure 10.4: The Non-enzymatic Synthesis of Haemoglobin A$_{1c}$ (HbA$_{1c}$)

A_{1c} therefore provides a sensitive index of the degree of long-term blood glucose control: the poorer the control, the longer the blood glucose level stays elevated and the greater the amount of haemoglobin A_{1c} that is formed.

The demonstration of increased amounts of haemoglobin A_{1c} in diabetics has led to the suggestion that many of the complications of diabetes may be related to increased non-enzymatic glycosylation of key cellular proteins. Such glycosylation might result in altered activity, solubility or half-life of the protein. Indeed, it has been shown that a high glucose concentration promotes the glycosylation of the lens protein a-crystallin and leads to an increased opacity of the protein matrix resembling that seen in diabetic cataracts. In addition, plasma proteins such as albumin and the apoproteins of low density lipoproteins are also found to be more highly glycosylated in diabetics than normal. Normally about 8 per cent of the albumin is glycosylated but this can increase to as much as 30 per cent in diabetes. Because of the relatively short half-life of albumin in the blood ($\sim$ 17 days) compared to that of haemoglobin ($\sim$ 120 days), measurement of glycosylated albumin may give an indication of the short-term control of blood glucose concentrations.

There is a 2-3-fold increase in the number of lysine residues glycosylated in the apoproteins of low density lipoproteins (LDL) from diabetics compared to those from non-diabetics. LDL are thought to play an important role in the development of atherosclerosis as there is a strong correlation between elevated LDL levels in plasma and the development of atherosclerosis both in diabetic and non-diabetic patients. The rapid clearance of LDL from plasma is dependent on a receptor-mediated endocytotic process involving the recognition of apolipoprotein B in the lipoprotein particle by high affinity receptors present on the surface of many cells including smooth muscle cells. The glycosylation of lysine groups on apolipoprotein B, the major apoprotein of LDL, interferes with the ability of LDL to bind to the high affinity receptor and thus inhibits the metabolism of LDL via the LDL-receptor pathway. The increased glycosylation of lysine residues in LDL from diabetics makes it likely that the metabolism of these particles is abnormal in diabetes and this might possibly play a role in the accelerated atherogenesis of the diabetic patient.

Diabetic microangiopathy results from a thickening of the capillary basement membrane due in part to an increase in the deposition of glycoprotein. The increased glycosylation of proteins observed during hyperglycaemia may be responsible for this excessive production of glycoproteins. In addition, the level of glycosylation of lysine and hydroxylysine residues in collagen isolated from the glomerular basement membrane has been shown to increase in diabetes. Since these residues participate in collagen cross-linking their increased glycosylation in diabetes has important implications with respect to the increased permeability of the glomerular basement membrane that occurs in diabetic nephropathy. Increased glycosylation of the lysine residues of proteins present in peripheral nerves has also been shown to occur in diabetes and this may relate to the degeneration that characteristically develops in this tissue as a result of diabetes.

Thus it is becoming clear that hyperglycaemia produces an increase in the rate of glycosylation of many tissue proteins and in some instances this has been shown to alter the properties of the protein. Increased glycosylation of key proteins in tissues involved in the complications of diabetes has been demonstrated and it seems likely that the glycosylation may be responsible, at least in part, for the tissue abnormality. However, it has yet to be shown that increased glycosylation of proteins leads directly to the development of the complications of diabetes. Such a theory could however account for the apparent relationship between the degree of hyperglycaemia and the extent of the complications since the rate of glycosylation is proportional to the glucose concentration. In addition, it would account for the fact that many of the complications of diabetes appear to result from abnormalities in the vascular system and this is the system most directly affected by hyperglycaemia.

Summary

It is now rare for a diabetic to die as a direct result of the acute metabolic derangements of the disorder since these are related to a relative or absolute insulin deficiency which can be corrected by insulin injection. However, the chronic complications of the disease are still a major clinical problem and this has raised the question as to whether the complications, like the metabolic disturbances, are consequences of an underlying tissue defect, or whether they arise as a consequence of the metabolic disturbances.

It appears likely that many, if not all, of the complications of diabetes arise as a result of the metabolic disturbances of the disorder. Thus for the most part the clinical complications, unlike the metabolic disturbances, are not present at the onset of diabetes in young people, but develop with increasing frequency as the duration of the diabetes lengthens. In addition, poor control of the blood glucose appears to predispose the diabetic to microvascular complications and neuropathy. Furthermore, the complications of diabetic pregnancy are largely corrected by strict metabolic control throughout the entire duration of the pregnancy. However, not all long-standing diabetics develop these problems and genetic and environmental factors may also be important.

It seems likely that diabetes arises, in certain individuals, as a result of a complex interaction between the individual and the environment. This interaction results either in the loss of functional B-cells or in an impaired function of the insulin secretory mechanism. In all cases it appears likely that the relative or absolute insulin deficiency is the major factor in the development of the metabolic abnormalities of the disorder and that failure to correct these abnormalities leads to the chronic clinical complications of the disease. Thus it should be the aim of all diabetic therapy to ensure that blood metabolite levels are maintained within the appropriate physiological range at all times.

Further Reading

Brownlee, M. & Cerami, A. The Biochemistry of the Complications of Diabetes Mellitus. *Annual Review of Biochemistry* (1981) *50*, 385-432

Bunn, H.F. Evaluation of Glycosylated Haemoglobin in Diabetic Patients. *Diabetes* (1981) *30*, 613-617

Cerami, A. & Koening, R.J. Haemoglobin A_{1c} as a Model for the Development of the Sequelae of Diabetes Mellitus. *Trends in Biochemical Sciences* (1978) *3*, 73-75

Clements, R.S. Diabetic Neuropathy – New Concepts of its Aetiology. *Diabetes* (1979) *28*, 604-611

Hosking, D.J. *et al.* Diabetic Autonomic Neuropathy. *Diabetes* (1978) *27*, 1043-1053

Palmberg, P.F. Diabetic Retinopathy. *Diabetes* (1977) *26*, 703-708

Simpson, L.O. Further Views on the Basement Membrane Controversy. *Diabetologia* (1981) *21*, 517-519

Sidenius, P. The Axonopathy of Diabetic Neuropathy. *Diabetes* (1982) *31*, 356-363

Spiro, R.G. Search for a Biochemical Basis of Diabetic Microangiopathy. *Diabetologia* (1976) *12*, 1-14

Steiner, G. Diabetes and Atherosclerosis. *Diabetes* (1981) *30 suppl. 2*, 1-7

Stout, R.W. Diabetes and Atherosclerosis – The Role of Insulin. *Diabetologia* (1979) *16*, 141-150

INDEX

A-cell 16-18
actin 18-20, 58
A-granules 17, 87
alloxan 147
angiopathy 155-8
artificial B-cell 139-40
atherosclerosis 155-6

B-cell 16-24
  ultrastructure 18-24
B-granules 16-17, 21-2, 25, 29, 56-9
biguanides 135-6

calmodulin 48, 54-5, 81
cataracts 159-60, 165
cell coat 18
cell web 18
chlorpropamide-alcohol flushing 143-4,
  151
C-peptide 36, 39-40, 127, 130

D-cell 16-18
desmosomes 20
D-granules 17
diabetes
  aetiology 142-9; autoimmunity
    145-6; genetic studies 142-4;
    histocompatibility antigens
    132, 144-5; viruses 146-7
  complications; basement mem-
    brane thickening 158-9, 165-6;
    haemostasis abnormalities
    157-8; lipoprotein abnormalities
    156; macroangiopathy 155-6;
    microangiopathy 155, 157-8;
    nephropathy 153, 157-9, 166;
    neuropathy 153-4; platelet
    abnormalities 156-7; protein
    glycosylation 158, 160, 163-6;
    red blood cell abnormalities
    157-8; retinopathy 153, 157,
    159-60
  definition 132-3
  diagnosis 133-4
  pathogenesis 149-51

pregnancy 160-1
  role of glucagon 148-9
  role of somatostatin 149
  treatment 134-42; artificial B-cell
    139-40; diet therapy 134-5;
    drug therapy 135-6;
    exercise 140-2; human
    insulin 138-9; insulin therapy
    136-9; islet cell transplantation
    140
  type 1 11, 67, 127-9, 132-3, 149-51
  type 2 11, 67, 129-30, 132-3,
    149-51
dihydrosomatostatin 103

endocytosis 24-6, 63
exercise 92-4, 117, 122-3, 140-2
exocytosis 24-6, 59

fasting 92-4, 116, 120-2
feeding 116, 119-20
fructose 162-3

gap-junctions 20, 116
gestational diabetes 125, 151
glicentin 89-90
glucagon
  deficiency 101
  mechanism of action 95-101
  metabolism 91
  physiological actions 91-5, 116
  secretion 90-1
  storage 87-8
  structure 86-7
  synthesis 30, 88
glucagonoma 101, 127
glucose tolerance test 133
glutathione 84, 161
glycoproteins 158, 165-6
glycosylation 158, 162-6
  albumin 165
  haemoglobin 158, 163-5
  lens proteins 160, 165
  lipoproteins 156, 165

haemoglobin A_{1C} 158, 163-5

168